Hybrid IT and Intelligent Edge Solutions
TOOLS AND TECHNOLOGIES TO ACCELERATE DIGITAL TRANSFORMATION

Marty Poniatowski

HPE Press
660 4th Street, #802
San Francisco, CA 94107

Hybrid IT and Intelligent Edge Solutions:
Tools and technologies to accelerate digital transformation
Marty Poniatowski

© 2018 Hewlett Packard Enterprise Development LP.

Published by:

Hewlett Packard Enterprise Press
660 4th Street, #802
San Francisco, CA 94107

All rights reserved. No part of this book may be reproduced or transmitted in any form or by any means, electronic or mechanical, including photocopying, recording, or by any information storage and retrieval system, without written permission from the publisher, except for the inclusion of brief quotations in a review.

ISBN: 978-1-942741-92-3

WARNING AND DISCLAIMER
This book provides information about Hybrid IT and Intelligent Edge tools and technologies based on Hewlett Packard Enterprise's portfolio of products, services, and solutions. Every effort has been made to make this book as complete and as accurate as possible, but no warranty or fitness is implied.

The information is provided on an "as is" basis and is subject to change without notice. The author, and Hewlett Packard Enterprise Press, shall have neither liability nor responsibility to any person or entity with respect to any loss or damages arising from the information contained in this book or from the use of the discs or programs that may accompany it.

The opinions expressed in this book belong to the author and are not necessarily those of Hewlett Packard Enterprise Press.

Feedback Information

At HPE Press, our goal is to create in-depth reference books of the best quality and value. Each book is crafted with care and precision, undergoing rigorous development that involves the expertise of members from the professional technical community.

Readers' feedback is a continuation of the process. If you have any comments regarding how we could improve the quality of this book, or otherwise alter it to better suit your needs, you can contact us through email at hpepress@epac.com. Please make sure to include the book title and ISBN in your message.

We appreciate your feedback.

Publisher: Hewlett Packard Enterprise Press

HPE Press Program Manager: Michael Bishop

Foreword

Technology continues to advance at a lightning-like speed. To succeed, enterprises of the future will require intelligent and autonomous IT systems that seamlessly connect their data from edge to core to cloud. Digital transformation is disrupting every industry and customers are looking for a trusted partner to help them navigate this new world.

With our innovative mindset and technologies, Hewlett Packard Enterprise is better positioned than ever to enable customers and partners to take advantage of the opportunities that exist now and that will emerge in the near future. At HPE, we are focused on building the world's best infrastructure solutions – intelligent, state-of-the-art technologies and services that provide the tools to harvest, analyze and store critical data from across the business. By quickly turning data into insights, enterprises will drive new business models, create new customer and workplace experiences, and increase operational efficiency.

In this book, HPE experts describe a number of tried and tested solutions that demonstrate these principles of innovation and collaboration in action. They illustrate the fact that Hewlett Packard Enterprise offers not just one or two locked-down proprietary platforms but a whole suite of open and flexible IT solutions with innovations such as HPE Synergy, OneSphere, and Aruba purposely designed to bridge the evolution to software-defined composable infrastructures, multi-cloud environments, and IoT at the edge. This aligns with HPE's vision to make hybrid IT simple, power the intelligent edge, while providing the services and expertise to make it happen.

In the end, it's all about the right mix of technology plus the right customer experience, driving the right business outcomes. To go further, faster. To accelerate what's next on your transformation journey.

Best Regards,

Antonio Neri
Hewlett Packard Enterprise
Chief Executive Officer

Introduction

The world of technology is changing and, at Hewlett Packard Enterprise (HPE), we are helping lead that change. To transform your business, you care about cutting edge topics such as gaining actionable insights at the intelligent edge, evolving the right mix of on-premises and cloud-based solutions for hybrid IT, and taking advantage of emerging tools, technologies, and services such as consumption-based spending for IT. You came to the right place to learn about these, and many other topics, in *Hybrid IT and Intelligent Edge Solutions*.

At HPE we have the unique ability to both advance IT continuously in our services and product portfolio and, at the same time, develop truly breakthrough technologies such as memory-driven computing. The chapters in this book cover tremendous HPE innovations that you can use today, including:

- Aruba, a key enabler of the Intelligent Edge
- Gen10, providing uncompromising security for next generation compute
- Superdome Flex, a scale-up solution with advanced SGI technologies
- Hyperconverged with SimpliVity, the ideal platform for hosting VMs
- Synergy, a new category of composable infrastructure to simplify hybrid IT
- Software-defined OneSphere for streamlined and simplified multi-cloud management
- Nimble and 3PAR, industry leading all flash storage systems
- OneView and InfoSight, powerful tools to manage and optimize your IT infrastructure

On the revolutionary horizon is memory-driven computing with The Machine. This is a change in the way compute will take place with memory at the center to dramatically improve processing performance and capabilities. Another technology that has had a meteoric rise is Blockchain, and here we take a look at the ways HPE can help you leverage Blockchain innovations for business success.

This book also contains numerous examples that you can view and potentially use as a starting point for your immediate IT needs. The HANA and HPC chapters provide a blueprint for the way in which we solved a specific problem for a customer using our products, services, and design expertise. At the same time, this book does not claim to discuss every solution that HPE has to offer. Cloud Technology Partners (CTP), for example, is the arm of HPE that helps firms manage the many aspects of developing the right cloud strategy. So, although this a topic that merits its own chapter, we do not cover here how CTP can take the confusion out of developing a hybrid IT strategy by providing consulting design and operational advisory services.

Why we wrote this book

At HPE our services and product portfolio is expansive and no book could comprehensively cover everything that we have to offer. As evidence of this, we would like to reference our previous book, *Ideal Platforms for Optimizing IT Workloads* published in 2017. That book covers twenty more workload solutions tailored to specific environments, and is essentially a companion volume to this. This edition is likewise meant to serve as a starting point to explore and understand the power of HPE technology to solve problems and seize opportunities. We hope you will be rewarded with many practical insights into how these HPE solutions can not only help drive, but also accelerate, your business transformation.

No matter what workloads you're running there is the ability to run them in a Hybrid IT manner, using a combination of on-premises data centers, private and public cloud. Similarly, any asset that you possess has the ability to provide useful data starting with the Intelligent Edge or Internet of Things. These two broad areas are the focus of this book and important to every business that we speak to about their IT needs. However, these conversations rarely lead to "one size fits all solutions" and virtually everyone has a unique approach to addressing these technologies and, in many cases, are not yet fully engaged with implementing them or are in the early stages of their use. Take containers, for example. Every business has a plan to implement containers, both as a way to streamline DevOps and as a great alternative to virtualization, but not many organizations have the majority of their workloads running in containers. This will advance, of course, but it provides an illustration of the transitions that continue to unfold and to evolve and where HPE can act as your trusted advisor in mapping a pathway to success.

How this book is organized

This book is designed to provide insights on tools and technologies that are designed to help accelerate your digital transformation journey. Chapters can be read independently to dive deeper on topics of interest or to explore solutions that are directly connected with your business needs and strategy. A summary at the end of each chapter highlights essential takeaways related to the topic at hand. The following is a list of some important chapters with an overview of key topics discussed:

Intelligent Edge

Chapter 1 is written by an expert who leads a team working on Intelligent Edge and Internet of Things (IoT). This is an area that is just beginning to realize its promise and this chapter covers many factors related to a successful Intelligent Edge endeavor.

Chapter 2 also relates to the Intelligent Edge and focuses on Aruba-related technologies. A key aspect of Intelligent Edge is having a network in place that is setup in a mobile first manner to overcome the limitations inherent using a traditional static and rigid network architecture.

Software Defined Data Center (SDDC)

The concept of software defined affects every aspect of Hybrid IT, so it is important to look at this area ahead of chapters on specific technologies and solutions. Chapter 3 covers many topics related to SDDC including bimodal IT in which two IT delivery models need to be supported: stable and agile. Another topic of particular interest in this chapter is HPE OneSphere which is a private-public cloud dashboard for your Hybrid IT environment.

Gen10

Much of the advanced hardware technology covered in this book is based on HPE Gen10 servers. Chapter 4 covers the key areas of Gen10 advancements including security, intelligent system tuning, persistent memory, and processor choice. Gen10 is also part of many subsequent chapters including Synergy, SimpliVity, Superdome, and Azure Stack to name a few.

Azure Stack, SimpliVity, Synergy, HANA TDI, Superdome Flex and HPC

The next set of chapters (5, 6, 7, 8, 10, and 12) focus on some typical customer problems to show you how a solution can be implemented. They provide some specific examples of how digital transformation can be achieved using the advanced capabilities of HPE's Gen10 and other platforms. There are, for instance, scale-up applications such as HANA TDI that requires as much memory as you can throw at them, and High Performance Computing (HPC) problems that can be solved with ingenious scale-out technologies. Some problems can be solved on a single platform, such as Synergy, which can be used for both scale-up and scale-out.

Memory-Driven Compute

In the age of big data, the information we are collecting is skyrocketing and we need memory storage that can process vast amounts of data, quickly, to extract business value. HPE Labs has developed a memory-driven computing model with The Machine that has, at its center, non-volatile memory, with System on a Chip (SoC) accessing this large pool of non-volatile memory. All of this technology is connected with a photonic fabric. This topic is covered fully in Chapter 9.

HPE OneView

The management of much of the modern IT infrastructure discussed in the preceding chapters can be done using HPE OneView. Chapter 11 discusses how OneView can not only handle basic management tasks on HPE products, such as BIOS and firmware updates, but can also be used to perform automation and orchestration on other platforms leveraging a powerful library of RESTful APIs.

Storage

The vast amount of storage required to support Intelligent Edge and Hybrid IT initiatives is growing dramatically. The need to process data in a high-performance and low-latency environment is key. In Chapter 13 on HPE Nimble, a section is devoted to InfoSight which helps you gain a meaningful understanding of application workloads by identifying I/O patterns, recurring performance patterns and noisy neighbor workloads. Storage Area Networks (SAN) and 3PAR solutions are covered in Chapter 14.

New IT Consumption Models and Secure Technologies

There are other chapters that cover a variety of topics including an alternative way to pay for your technology based on consumption (Chapter 15), and an overview of Blockchain (Chapter 17) which is quickly emerging as a disruptive technology in many industries. There is also a chapter on using secure encryption technologies for added security (Chapter 16).

In summary, this book is a window into Hybrid IT and Intelligent Edge from the perspective of both reference architectures as well as real-world proof points and design examples. I hope that you will find many useful pointers on ways to plan and execute your IT strategy using the best tools, technologies, and services to accelerate your digital and business transformation.

Thank you,

Marty Poniatowski
marty.poniatowski@hpe.com

About the Authors

This book consists of submissions from many authors who crafted the solution covered in their area of expertise. The primary author worked closely with the chapter authors to produce a book that comes as close as possible to reading as if one author produced the entire book.

Primary Author

Marty Poniatowski is a Senior Director at Hewlett Packard Enterprise managing the presales technical experts in the U.S. His team is comprised of Account Chief Technologists and Enterprise Architects who focus on the key initiatives of their clients and craft solutions to these complex problems. In so doing, his team works not only with HPE services and products but also firms with which HPE has partnerships.

During his career Marty has authored 19 books on IT topics. Marty holds graduate degrees in Information Technology (New York University) and Management, and an undergraduate degree in Electrical Engineering.

Chapter Authors

The following is a list of authors in alphabetical order:

Gary Allard is a Senior Solution Architect at HPE focused on server and storage solutions in the state government, local government, education, and healthcare industries.

Hersey Cartwright is an HPE SimpliVity Hybrid IT presales Solutions Architect and is a VMware Certified Design Expert (VCDX). Hersey is focused on datacenter virtualization and hyperconverged infrastructure.

Christian Crisan is an Account Chief Technologist focusing on some of the largest Financial Services Industry Clients. Christian is driving the Pan-HPE and partner technical strategy centered on business to technology alignment and emerging technology trends.

Larry Fondacaro is Manager of the HPE East Enterprise Architecture (EA) team that provides technical solutions across all industry verticals to HPE enterprise customers.

David J Furtzaig is a Financial Services Industry Chief Technical Strategist and Technologist, focusing on Public/Private/Hybrid Clouds, large-scale enterprise solutions and technology simplification.

Tom Golway is a Chief Technologist at HPE focused on providing business and technical thought leadership to help customers realize their business strategies. Tom's primary focus has been in working with customers on innovative use cases for emerging technologies such as Blockchain, Deep Learning, and Memory Driven Computing.

Steve Gorgone is a senior Solutions Architect in HPE's Hybrid IT group and is focused on delivering solutions around HPE's Synergy, SimpliVity and OneView.

Michelle Carlson Hannula is a senior marketing leader focused on HPE's software defined and cloud technologies. Michelle's background includes expertise across hardware, software and virtualization technologies.

Nick Harders manages a team of Aruba Systems Engineers (SEs) covering the New York Metro region and has a personal interest in network programmability and orchestration.

Zvadia Hibshoosh is an HPE Solution Architect supporting large enterprise accounts, primarily in the Financial Services Industry.

Patrick Hilley is a Solution Architect for HPE's Server Technologies. Patrick's current focus is HPE's Hybrid IT solutions including HPE Synergy, HPE SimpliVity, and HPE OneView.

Hossein Hosseini is an Account Chief Technologist, strategist, evangelist, and change agent who builds technical trusted advisor relationships with CIO/CTO, SVP, Director, and Chief Technical Leads.

Khalil Hraiche is an Enterprise Architect at Hewlett Packard Enterprise focused on the Financial Services Industry. Khalil represents the complete HPE portfolio and works directly with HPE Global and Enterprise customers.

Dave McNamare is an experienced SAP Solution Architect for HPE's Mission Critical Server Technologies focused on sizing and infrastructure solutions for SAP workloads on HPE's SAP HANA Converged Systems, SAP TDI HANA Servers, Mission Critical X86 Servers, and Integrity HP-UX Server solutions.

David Robert assists enterprise customers to design and build infrastructure solutions for a wide variety of applications and workloads, most recently focusing on HPE's industry standard server product lines.

Chuck Strobel is an experienced Master Solution Architect for HPE's Mission Critical Server technologies. Chuck is focused on HPE's SAP HANA Converged Systems, Mission Critical x86 Servers, and Integrity HP-UX Server solutions.

Nick Triantos is a Field CTO and Storage Architect for HPE's Storage technologies. Nick is focused on Nimble Storage, HPE Cloud Volumes and Predictive Analytics.

John Tsang leads the IoT and Big Data business for HPE North America sales. His responsibilities include the IoT go-to-market and operations readiness to develop the PAN HPE IoT business. John is executing new PAN HPE IoT solutions and initiatives to help customers transform and take advantage of this tremendous technology revolution underway.

Vito Vultaggio is an experienced technology leader at HPE in the enterprise storage Business Unit. Vito is focused on all aspects of Storage technologies, specializing in SAN, NAS, scale out architectures, and integration of our storage portfolio into our Hybrid IT strategy.

Robert Wallos is a technology strategist who has worked for and with some of the largest financial institutions in the world. As Chief Technologist focusing on the Financial Services Industry, he aligns HPE's technology strategy with that of some of HPE's largest financial clients.

Joseph Yanushpolsky is an Enterprise Solutions Architect with HPE Northeast team specializing in design, sizing, and tuning of complex infrastructure and application environments.

CONTENTS

1 The Internet of Things (IoT) and the Intelligent Edge 1
   The Internet of Things 1
      Big Data Analytics and Machine Learning... 2
      The Evolution of Computing 2
   The Rise of the Intelligent Edge 3
      Benefits of Compute at the Edge 4
      Impacts of Real-World Information 6
      Three "Cs" of the Intelligent Edge 7
      A Business Revolution: Digital Disruption at the Edge 8
   HPE Converged Edged Systems 8
      IoT Consolidated Architecture 9
   Summary 11

2 Aruba Mobile First Campus Architecture 13
   Solution overview 13
   Mobile First Access—Traditional reference design 14
      Access components 14
   Mobile First Access—Centralized reference design 18
      Access components 19
      Dynamic segmentation—Use cases and scaling considerations 19
   Mobile First Access—Distributed reference design 20
      Access components 21
   Edge optimization for Converged Wired + Wireless 22
      AP integration 22
      ClearPass integration 23
      AirWave integration 24
      Zero-Touch Provisioning (ZTP) 24
   Mobile First Backbone—Collapsed and hierarchical reference designs 25
      Backbone components and virtual switching framework 26
   Sample reference design—Single building campus 29
      Infrastructure components 30
      Technologies and protocols 30
   Sample reference design—Multiple-building campus 30
      Infrastructure components 31
      Technologies and protocols 31

Single VLAN architecture for wireless clients..31
Summary..32
    References 32

3 Software Defined Data Center (SDDC) 33
    Digital Transformation in the Hybrid IT journey..................................33
    The Software-Defined Data Center (SDDC)..34

HPE OneSphere 35
    Enabling IT, LOB, and developers collaboration................................35
    Accelerating digital transformation with bimodal IT 36
    Example 1: SimpliVity microservices orchestration.............................37
    Example 2: Synergy microservices and traditional composability........38
    Using OneSphere to build, operate, and manage hybrid IT 39

HPE consumption-based services 39
    HPE GreenLake solutions 40
    Professional services 40

Summary..41

4 HPE ProLiant Gen10 Advancements 43
Enhanced security 44
    Protect, detect, and recover..44
    Additional security options..46

Gen10 memory enhancements...47
    Persistent memory NV-DIMM 47
    Scalable persistent memory 48

Intelligent System Tuning (IST)..48
    Jitter smoothing 48
    Core boosting...49
    Workload matching 49

Higher levels of compute performance 50
    Enhanced processors..50
    Support for Graphical Processing Units..50

Increased in-server storage density 51
More efficient and easier server management 51
Example of Gen10 workloads enhancements 52
    Solution overview 53

Summary..55

5 The Power of Azure in your Data Center: HPE ProLiant for Microsoft Azure Stack 57
Microsoft's Hybrid Cloud Platform 57
Solution overview 58
Hardware inventory 59
Key use cases 61
Ensure compliance, data sovereignty, and security 61
Maximize performance 61
Connect edge and disconnected applications 62
Accelerate modern application development 62
Summary 62

6 HPE SimpliVity Hyperconverged Infrastructure 65
Solution overview 65
Solution components 66
Architectural features and benefits 67
Storage efficiency 67
Configurable compute resources 69
Scale-out workloads 69
Compute-only scaling 70
Resiliency 71
Hardware inventory 71
Summary 73

7 HPE Synergy and Composable Infrastructure 75
Why Synergy? 75
Synergy overview 76
HPE composable infrastructure 77
Synergy 12000 frame and components 78
Solution architecture 85
Hardware components 85
Summary 88

8 HPE HANA TDI Solution and Superdome Flex 89
Solution requirements 89
Business requirements 89
Technical requirements 90
Solution overview 90
How did we arrive at this solution? 92
Hardware inventory 96

Software inventory 97
Implementation plan..99
Solution validation and success criteria 101
    SoH Testing 101
    Deployment overview...101
HPE Superdome Flex 102
    Example Superdome flex implementation......................................103

9 Memory-Driven Computing..107
The evolution of IT consumption models..107
    System of Record 108
    System of Engagement 108
    System of Action 108
    New processing paradigms 109
    Trends in application architecture 109
    The end of scaling as usual 109
    The data revolution 111
Memory-driven computing (MDC) 112
    Core components..113
MDC software architecture..115
    MDC software stack 116
    MDC developer toolkit...116
Workload categories and examples...119
    MDC time machine 119
    Transportation and logistics 120
    Financial services..122
    MDC Enterprise Service Bus (ESB) 122
    MDC similarity search 123
    HPE Moonshot and System on a Chip (Soc)..................................124
Summary...125
Additional resources...126

10 Worldwide VDI Deployment on HPE SimpliVity 127
Solution requirements 127
Solution overview...128
    Solution details 130
    Rack Power requirements...135
    Solution deployment...136
Summary...137

11 OneView Integration 139
RESTful-based industry solutions 140
How HPE OneView works 140
HPE Image Streamer 140
A single unified workflow 141
Example of HPE OneView integration with Ansible 141
Example of HPE OneView used with Docker 144
Summary 145

12 High Performance Computing 147
Solution overview 148
Hardware components 149
How did we arrive at this solution? 150
Hardware inventory 153
Implementation project plan 156
Summary 156

13 HPE Nimble: Mixed Virtualized Workloads 157
Design considerations 157
Solution requirements 158
Solution overview 159
Scaling capabilities 161
Management capabilities 161
Advanced analytics 162
InfoSight predictive analytics 163
Hardware inventory 166
Summary 167

14 High-Performance SAN and All-Flash 169
SAN considerations and overview 170
Storage 170
Storage Area Network (SAN) 171
NVMe over Fiber Channel (NVMe-oFC) 172
Compute 173
Server Host Bus Adapter (HBA) 173
Applications 174
SAN extension 174
Management Tools 176
HPE SmartSAN for 3PAR StorServ 176
Brocade EZSwitchSetup 176
Brocade Fabric Vision 176

Small, Medium, and Large SAN 177
    Small, medium, and large SAN inventory 178
    HPE SAN Deployment Services 179
Summary 179

15 Consumption-Based IT Services 181
    What is consumption-based IT? 181
    What are its benefits and advantages? 182
HPE GreenLake: Pay-per-use IT solutions 183
    Consumption-as-a-Service 184
    Flexible capacity 184
    HPE GreenLake portfolio 185
Sample design and implementation 187
    Hardware and software components 189
    HPE Datacenter Care Service 190
    Phased implementation approach 190
Summary 191
    Why choose HPE? 191
    Additional resources 192

16 HPE Secure Encryption 193
Why This Solution? 193
    Broad Encryption Coverage 193
    Flexibility 194
    High availability and scalability 194
    Simplified deployment and management 194
    Regulatory compliance 194
Solution overview 195
    Key management 196
    Encryption algorithms 196
    Enterprise Secure Key Manager (ESKM) 197
    Licensing 198
    Considerations 199
    Competitive view 199
System Management 200
Related solutions 200
Summary 200
    Additional resources 201

17 Blockchain Demystified 203
Blockchain fundamentals 204
Cryptography and crypto currencies 205
Blockchain frameworks..207
Blockchain: Public, private, and consortium 208
Blockchain: Use cases 209
Smart contracts..210
Smart assets 211
Clearing and Settlement 212
Payments..213
Digital identity 215
Blockchain challenges..216
What are the headwinds facing this new Technology?.....................216
When Smart Contracts are not so smart..219
When the distributed ledger is public..219
Lack of common architectures 219
Hewlett Packard Enterprise solutions 220
Distributed ledger usecases for financial services 220
Summary..224

Index 225

1 The Internet of Things (IoT) and the Intelligent Edge

INTRODUCTION

The new technology revolution is here, and for the first time in history, technology has exceeded humankind's ability to use it and has provided an "Art of the Possible" for leveraging information technology for almost anything. The combinations of cloud computing, mobile, and connected everything, everywhere and at any time along with the ability to sensorize and digitize any objects and things have produced an era of digital and connective disruption the like of which we have never seen before. This chapter discusses the possibilities presented by HPE technology for powering the new frontiers at the Intelligent Edge.

The Internet of Things

No longer is science fiction just fiction, it is becoming reality at a furious pace and what is feeding all this is the Internet of Things and for Things (IoT). IoT is feeding this wave of the new big data where old analog data that have long existed everywhere is now being captured and digitized to gain faster time to insights and real-time actions. Figure 1-1 depicts the type of objects that can be connected in IoT.

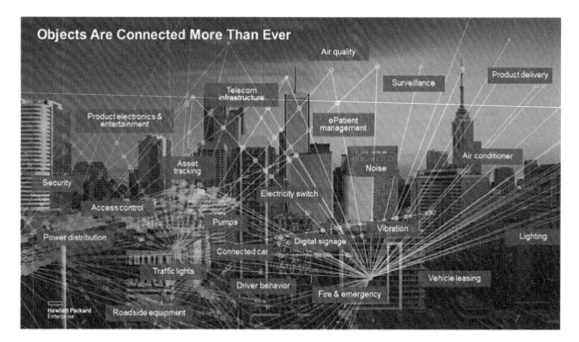

Figure 1-1 Objects and IoT

Big Data Analytics and Machine Learning

With all of these objects connected, there is a tremendous amount of data being collected. This new big data is now the new "gold," and the ability to mine data for intelligence from anything is in turn enabling a whole new era of distributed analytics and edge intelligence. The information loop—of being able to sense the data, infer the data, and take action on the data—allows for distance deep learning from the edge to core to cloud and back, creating unlimited number of algorithms for unlimited outcomes. From this, the ability for machine learning and ultimately true Artificial Intelligence (AI) will influence all aspects of life as we know it. These technology buzzwords are now becoming household vocabulary, and we are entering the era of what may be all things possible.

The Evolution of Computing

In order to truly grasp the magnitude of this from a compute point of view, more clarity is needed to define how we got here and what this means for the future. Peter Levine, a general partner of Andreesen Horowitz, does a very effective job of describing the evolution of computing. As Figure 1-2 illustrates, the history of computing continues to be cyclical with centralization starting with the mainframes in the 1960s–1970s moving to a distributed client-server architecture with the advent of desktop and personal computers in the period from 1980 to 2000; then back to a centralized computing with the current mobile-cloud explosion, now accelerating back to a distributed model with edge intelligence. Along this timeline, the systems and devices have grown exponentially from ten million mainframes to multibillions of client-servers to innumerable machines in mobile cloud.

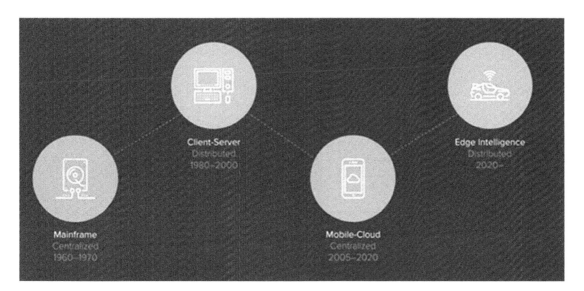

Figure 1-2 The Evolution of Computing
Source: Peter Levine

In intelligent edge, we are talking in terms of trillions of systems, devices, and machines due to the fact that in the age of the "Intelligent Edge" humans are not the common denominator but rather machines. Machines will be communicating to machines introducing the potential for amazing innovations such as autonomous vehicles, cognitive computing with robotics, and smart everything whereby everyday objects, devices, things and machines communicate seamlessly with one another to improve our everyday lives.

The Rise of the Intelligent Edge

Given the rise of the intelligent edge, what will happen to cloud computing? Will the cloud diminish or ultimately get eaten up by the edge? Using a sports analogy, the intelligent edge is the new playing field. The playbook and plays within the Intelligent Edge are virtually limitless creating and driving new business outcomes, use cases and ultimately changing the world. So let us define what and where the intelligent edge is, and how and why we want to enable intelligence at the edge and what will happen to the role of the cloud. At its most basic, the edge is anything outside the data center and/or cloud while the cloud is really a bunch of data centers that reside in various places. Figure 1-3 illustrates the key elements of the edge.

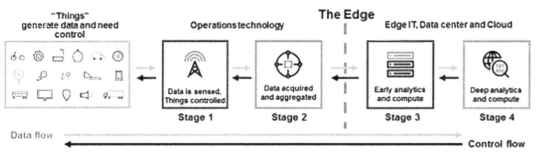

Figure 1-3 Defining the "Edge"

You can see that the edge has various levels, but simply put it is a place outside of the traditional data center or cloud, near to the things and, in many cases, mission-critical things such as operations technology (a spectrum of objects and systems that produce products and/or services). This convergence of information technology and operations technology in IoT is creating the intelligent edge and every industry, organization, and business is in an arms race to enable intelligence at the edge to create innovation, to deliver new business models, and to lower operation costs while gaining competitive advantages in the marketplace.

Benefits of Compute at the Edge

One primary business reason for distributing compute and doing analytics closer to the data source is the sheer volume, velocity, and variety of data. Many of today's desired business improvements can only be achieved by analyzing data that is being created at the edge or outside the data center. This data—that may be structured, semistructured, or unstructured—is growing exponentially in volume (megabytes to petabytes per day), variety (video, images, etc.) and velocity (real time). Considering the data economics, trafficking this data to a core/cloud data center for processing is cost-prohibitive and exposes the data to cyber-attack; thus a good deal of the data is going unanalyzed. Figure 1-4 highlights the benefits of performing computation at the edge.

Benefits of Compute at the Edge:
(and not send the edge data to the data center / cloud)

Figure 1-4 Benefits of Compute-At-The-Edge

The following list describes each benefit in detail:

1) **Latency:** Latency in data transfer reduces "time-to-insight" from the data, which slows "time-to-action" for businesses and protracted responses to the data. By processing the data at the point of ingest, the time to insight is substantially reduced because the data is processed in real time.

2) **Bandwidth:** Reduce dependency on the Wide Area Network (WAN) by processing all the data at the edge and only sending the anomalies and insights over the WAN. One of the biggest inhibitors to processing more of the data being generated is network limitations. Very Small Aperture Terminal (VSAT) is low bandwidth and unreliable. Third-Generation Partnership Project (3GPP) has greater bandwidth than VSAT but has its challenges as well. There may not be Ethernet in many remote locations, and if there is, it may not be the necessary high bandwidth or reliable. By using a LAN for transporting the data and WAN for transporting the anomalies and insights, companies can save a substantial amount of money on their network.

3) **Cost:** Sending data synchronously to a data center for processing can be expensive. Costs can be greatly reduced by containing the data at the edge and computing the analytics close to the source.

4) **Security:** Transferring data over a network by definition exposes data to security threats. By collecting and processing the data at the edge, data traffic and exposure is contained with only the anomalies and insights being sent over the WAN.

5) **Supportability:** Most edge locations lack IT staff, so you need a solution that can be supported physically by local staff and remotely by Information Technology (IT) or Operational Technology (OT.) A converged edge system can do what a smart phone has done with all the disparate devices and simplify use, manageability and support.

6) **Duplication and Durability:** Most edge locations do not have IT conditions or resources, so the solution used at the edge has to be durable and flexible. The ability to handle inclement conditions yet flexible enough to place just about anywhere like on a wall, under desk in a closet, or even in a truck.

7) **Compliance:** Data sovereignty with region and country compliance issues can complicate data transfer across borders and long distances. By processing the data at the edge, the data never has to leave the country.

Impacts of Real-World Information

For the first time, we have real-world information requiring real-time action. The challenges are exacerbated by limitations in the legacy technologies currently in place and the data being isolated or trapped in a variety of disparate repositories, often owned by a variety of different users, as shown in Figure 1-5:

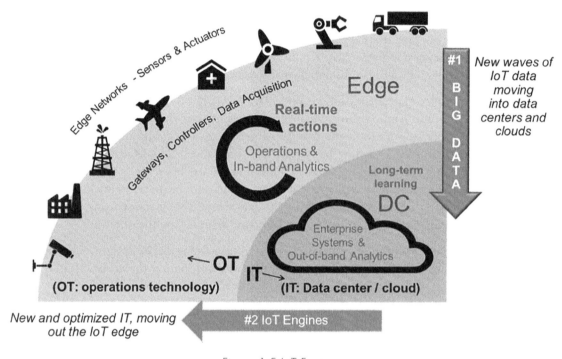

Figure 1-5 IoT Factors

Figure 1-5 illustrates that all the physical things at the edge, along with all the spectrum of systems that create products and services (many of which can be mission-critical), are being digitalized. This in turn creates vast amount of data that is either good (data that is useful immediately) or bad (data that cannot be used immediately but may have some purpose later) or ugly (data that can only be made useful with further analysis). The good data can drive immediate insights at the edge. The bad and/or ugly data, however, must typically be sent to the data center/cloud for long-term learning and data mining.

So the role of cloud does not really diminish but instead becomes a "training center" for data mining and creation of new algorithms that can then be propagated back to the end points for more decision-making. At the same time, data center compute capabilities are shifting left and closer to the vast operational systems and devices where all this data is originating. This is enabling many IoT software platforms and purpose-built analytic engines to be closer to the device level where it matters most and to maximize actionable data in real time. There is also a shift happening with cloud platforms and applications being extended to the edge and being consumed as a service. Powering the intelligent edge thus involves establishing a strong connectivity foundation to enable edge computing and enhanced control of systems and machines.

Three "Cs" of the Intelligent Edge

The 3Cs of the Intelligent Edge (connect, compute, and control) are shown in Figure 1-6.

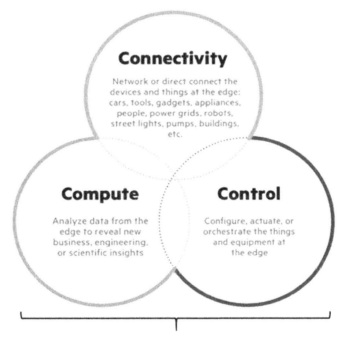

Figure 1-6 "The Three Cs"

The edge cannot become intelligent unless you have the right mix of the 3Cs, and it all starts with ensuring the hardware infrastructure is modernized from edge to core to cloud to take on the data and application workloads to deliver the IoT uses cases involved with driving the desired business outcomes.

A Business Revolution: Digital Disruption at the Edge

Enabling intelligence at the edge has a direct correlation to the digital disruption or transformation organizations and businesses are facing today. Companies are facing disruption to their business by competitors they never knew existed or would one day be their biggest competition. Who would have thought that Uber would be the leader in transportation without owning a single vehicle, or that Amazon, from their roots as online book e-tailer, would become the largest cloud service provider and now also venturing into the grocery and healthcare business?

The number one fear with companies today is that they cannot implement a data analytics strategy fast enough. Those that do, can have tremendous market advantages. As mentioned earlier, the "Art of the Possible" means that there is enormous value in creating new business models at a much faster pace than ever before, improving operational efficiencies and enhancing customer experiences. New technology waves are being created from the enablement of self-driving trucks to paying with your face using visual recognition software. Especially within industrial IoT and proven smart manufacturing use cases, users are able to monetize IoT benefits through significant reduction in downtime, improved quality and safety, and increased returns on assets. Figure 1-7 depicts some startling parameters related to IoT growth and investments.

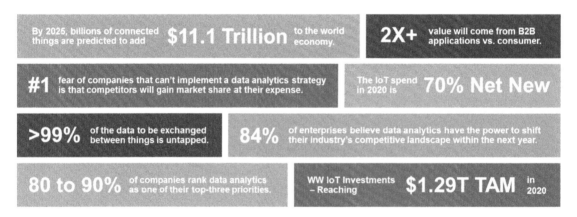

Figure 1-7 Parameters Related to IoT

HPE Converged Edged Systems

As businesses face transforming how they operate to be more competitive and profitable in a global, digital, economy, they are finding IT to be a key enabler to helping their business operations or operations technology (OT) achieve better business outcomes. In the past where IT, OT, and Line of Businesses (LoB) worked separately, now with this digital and connectivity revolution, there must be strong collaboration with IT and operations. To help overcome these challenges and bringing IT and

OT systems together, HPE developed the industry's first converged edge systems that can work in conjunction with the customers legacy assets to bring all the data together for a more holistic view of their operations, efficiently, by having access, connectivity, and compute closer to the data source. This distributed approach, using IT technology, allows businesses to better optimize their analytic needs to their budget and existing IT and OT environments thus minimizing disruption to existing operations. Figure 1-8 depicts these edge systems.

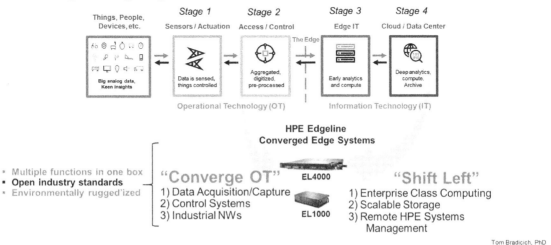

Figure 1-8 HPE Converged Edge Systems

IoT Consolidated Architecture

Along with this first mover innovation and differentiation with converged edge systems, HPE's Aruba networking platforms completes the 3Cs with the mobile first portfolio enabling intelligent spaces with location-based services, connect and protect with the Clearpass and Niara security platforms. No longer does the users follow the network, the network follows the users and secures all the disparate IoT devices no matter the industry. In addition, HPE's Universal IoT platform provides a carrier grade device and connectivity, which goes beyond WiFi, management and data acquisition platform scalable to handle the largest smart cities implementations.

Figure 1-9 depicts this IoT consolidated architecture.

HPE's Consolidated IoT Architecture

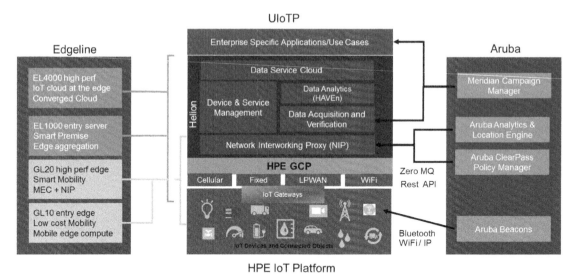

Figure 1-9 Consolidated IoT Architecture

There is an information loop of sense, infer, act, and machine learning (an early form of artificial intelligence) that unlocks the power of the edge. Similar to solving a multifaceted Rubik's cube using countless algorithms, the "Age of Algorithmic Businesses" is here in which organizations will be valued not only for their big data but also for the algorithms that turn that data into actions, resulting in an improving of the customer experience and increase of customer impact (Figure 1-10).

Figure 1-10 13×13 Rubik's Cube

For example, new technologies such as Blockchain, a peer-to-peer network built on top of the Internet, is enabled by this new wave of distributing, computing architecture formulating a distributed ledger system to automate trust in creation of new crypto currencies and peer-to-peer transactions. These new data types drive visibility and transparency into processes, help secure organizations against attack, boost the productivity of human and capital assets, and help drive profitability by identifying new business opportunities. Think in terms of an algorithm app store for business transformation. Based on specific use cases and the largely untapped capabilities of machine learning and artificial intelligence, there will be a new set of algorithms to deliver unlimited outcomes and as yet unimagined benefits to humankind.

Summary

The power of IoT and the Intelligent Edge comes from extracting process, business, and customer data that are locked inside enterprises, inside devices, inside machines, and inside infrastructure. Accelerating intelligence at the edge serves a number of goals from improved quality and reduced costs to increased responsiveness and improved time-to-market throughout the supply chain. Leading the charge, HPE's Converged Edge Systems and Consolidated IoT Architecture enables businesses to start unlocking the tremendous opportunities for IT and business transformation today.

2 Aruba Mobile First Campus Architecture

INTRODUCTION

This chapter describes the Aruba Mobile First architecture for campus networks, including reference designs, technologies, and hardware and software components. Aruba understands that rigid, static enterprise networks do not satisfy the needs of today's enterprises. Users move, have more than one device, access video and audio interactive applications as much as they access databases and use office applications.

The Aruba Mobile First network is designed to allow people to move while connected and to allow enterprises to innovate without being tied to legacy network infrastructure. It combines the best-of-breed wireless product line, a wired infrastructure ready to support mobility and IoT, end-to-end network management, and multivendor access control. This chapter illustrates the flexible design choices available when selecting this architecture for campus networks.

Solution overview

Historically, documents on campus network architectures describe wired and wireless networks as separate entities, one running on top of the other but with minor or no interaction between them. Aruba takes a different approach; we describe access as a single system that integrates APs, mobility controllers, access switches, network management, and access and traffic control.

For **Mobile First Access**, three reference designs are presented:

- **Traditional Reference Design** can be used in a vast majority of the cases, integrating controller-based wireless access and local forwarding access switching.
- **Centralized Reference Design** in which wireless and wired traffic is tunneled to the mobility controllers to completely unify policy enforcement.
- **Distributed Reference Design** in which traffic from wireless and wired clients is forwarded locally at the access layer using Aruba Instant technology.

For the network backbone (the switching system that transports all the traffic between the access system and the services and applications), two **Mobile First Backbone** designs are presented:

- **Collapsed Reference Design** is a single node (collapsed core) backbone design mainly used for single building and small campuses.
- **Hierarchical Reference Design** (with core and aggregation nodes) is a hierarchical backbone design typically used in multibuilding campuses.

Mobile First Access—Traditional reference design

The most common campus access design is based on a traditional access switching infrastructure and a controller-based mobility system. This design satisfies the requirements of most campuses and can scale to tens of thousands of clients (Figure 2-1).

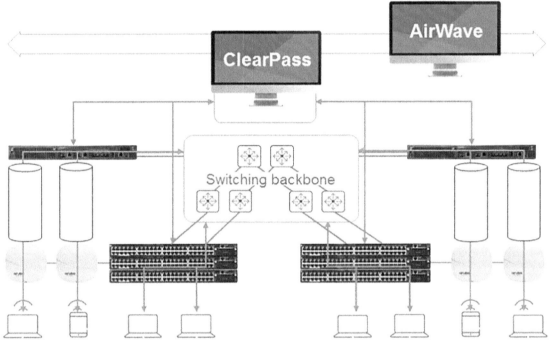

Figure 2-1 Traditional access reference design

Access components

The components for this design are Aruba APs, Aruba access switches, Aruba Mobility Controllers, AirWave, and ClearPass Policy Manager. These hardware and software components form an integrated access system.

Aruba access points (APs)

Aruba APs provide connectivity to wireless clients and other endpoints (Figure 2-2).

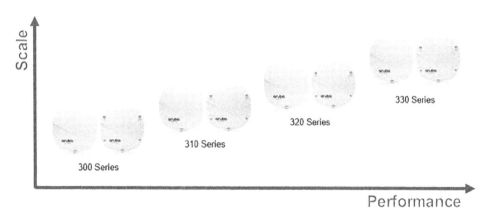

Figure 2-2 Aruba 802.11ac Wave 2 AP product line

Aruba access switches

Aruba access switches connect APs, wired clients, and other endpoints such as printers, cameras to the network. These switches on their own, or working with ClearPass Policy Manager, provide an access control service including authentication, authorization, and accounting (Figure 2-3).

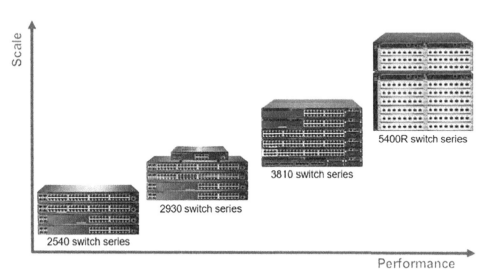

Figure 2-3 Aruba access switch product line

Aruba mobility controllers

Aruba mobility controllers perform different functions for the APs. Several controllers can work together to provide a hierarchical and redundant mobility system with the Aps (Figures 2-4 and 2-5).

CHAPTER 2
Aruba Mobile First Campus Architecture

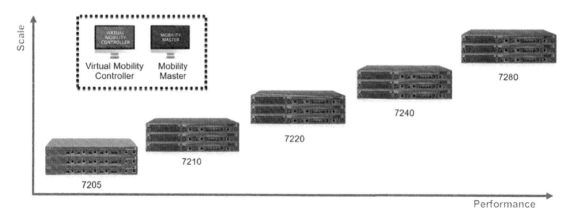

Figure 2-4 Aruba campus mobility controller product line

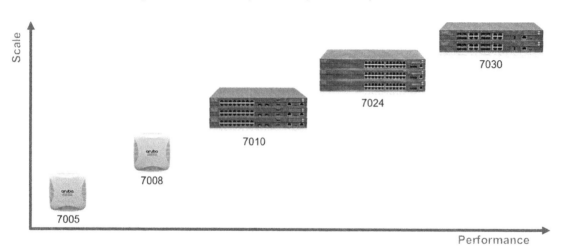

Figure 2-5 Aruba branch mobility controller product line

The mobility controller system provides:

- AP tunnel termination and translational bridging
  - Each AP establishes a GRE tunnel to a mobility controller.
  - All traffic to and from the AP's wireless clients is transported inside this tunnel: in other words, the controller provides a virtual connection point to all wireless clients.
  - Traffic arriving from a wireless client is translated from 802.11 to 802.3 frame format and traffic destined to a wireless client is translated from 802.3 to 802.11 frame format.
  - The frame translation function includes encryption and decryption of wireless traffic.
  - The controller can implement quality of service (QoS) by marking or remarking traffic and by providing traffic prioritization.

- Access control and policy enforcement
    - Working alone or in conjunction with ClearPass, the controller acts as the authenticator for wireless clients: in other words, it is the controller that rejects or authorizes mobile clients, and authorized clients are allowed to access only a certain set of services and resources.
    - The controller includes a stateful firewall that can be configured to filter wireless traffic including multicast and broadcast control.
- Network management and visibility
    - Aruba mobility controllers are deeply integrated with AirWave and together, these two platforms provide granular visibility into the wireless network, allowing the network administrator to detect and troubleshoot connection and performance issues.

Aruba AirWave

Aruba AirWave is a multivendor network management platform for wired and wireless networks. AirWave is especially suited to manage Aruba mobility including:

- The Aruba **AppRF** feature provides deep visibility into performance and usage of mobile and web apps. Reputation reports allow network administrators to quickly take action against high-risk sites and control Wi-Fi usage by app category and user role, device type, and location-specific insights, allowing them to make quick decisions to protect business-critical apps.
- AirWave uses **VisualRF** to allow for time-lapse mining of Wi-Fi coverage—enabling wireless network engineers to record and replay 24 hours of RF heat mapping. VisualRF also powers location-specific, visual analytics on mobile and web app usage, mobile device performance, mobile UC voice, video quality, and more.
- AirWave provides a **Clarity** dashboard to deliver insights into live user performance as well as synthetic testing within the Wi-Fi network. Clarity adds end-to-end visibility into the end-user experience, streamlining operational processes and troubleshooting activities.

AirWave also manages Aruba switches providing CPU, memory and interface monitoring, configuration management, firmware upgrades, and more.

AirWave can be used to implement zero-touch provisioning for Aruba Instant APs (IAP), Aruba switches, and Aruba branch mobility controllers.

Aruba ClearPass Policy Manager

Aruba ClearPass Policy Manager (CPPM) solves today's security challenges across any multivendor wired or wireless network by replacing outdated legacy AAA with context-aware policies. It delivers visibility, policy control, and workflow automation in one cohesive solution.

CPPM provides granular visibility into all connected devices, even controllers and switches including a customizable dashboard to display categories and families of devices and individual attributes for everything connected.

- **ClearPass Guest** allows network administrators to give customers, contractors, and other visitors secure guest access to wireless and wired networks including customizing a guest access portal.
- **ClearPass Onboard** lets BYOD and IT-issued devices connect safely to the corporate network in compliance with security mandates. Flexible policies and unique certificates enable full or limited access based on roles, device type, and security posture.
- **ClearPass OnGuard** performs vital endpoint health checks and posture assessments automatically to ensure that all laptops are fully compliant with industry and internal requirements before they connect to wired and wireless networks.

Mobile First Access—Centralized reference design

This second access design takes advantage of the mobility controllers located at the core to centralize not just wireless client traffic, but also wired client traffic (Figure 2-6).

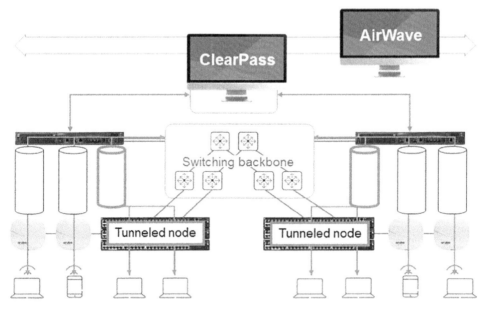

Figure 2-6 Centralized access reference design

Access components

The components and technologies for this design are the same as for the traditional approach except for the addition of an Aruba switch feature called **Dynamic Segmentation**, which is supported on the Aruba 5400R, 3810, and 2930 switch series, and all Aruba mobility controllers. Depending on the switch platform, Dynamic Segmentation can be deployed on a Per-Port Tunneled Node (PPTN) or Per-User Tunneled Node (PUTN) basis.

Dynamic segmentation—Use cases and scaling considerations

A switch or a stack of switches with dynamic segmentation enabled, creates Layer 2 GRE tunnels to a primary controller. A secondary controller can be configured for redundancy. Ports can be configured in tunneled "mode," and in those cases, all the traffic to and from these ports is encapsulated within GRE tunnels. For clients connected to these ports, the switch acts as an access point, and they share a virtual access port at the controller (Figure 2-7).

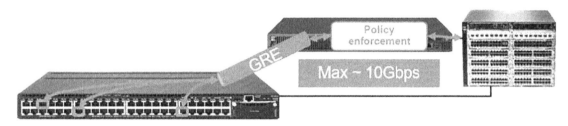

Figure 2-7 Per-port tunneled node

If all the ports of a switch or a stack are configured in tunneled mode, the whole switch behaves in a sense like an access point; however, because the clients are connecting via Ethernet, there are two main differences: first, there is no encryption on the tunnel, and second, there is no need for translational bridging by the controller.

It is important to note that in most cases only some ports will need to be configured in tunneled mode, while the remaining ports work as standard access ports. If using per-user tunneled node, differentiated access can be dynamically enforced for multiple devices connected to a single port.

A hybrid design is usually most appropriate—only traffic that would benefit by terminating on a controller is tunneled, with all other traffic switched locally (with an enforcement profile defined at the switch port).

There are two main use cases for dynamic segmentation:

Critical client protection: Some clients need to be protected from potential intruders, and using the mobility controller's firewall is a simple way to do it. By using dynamic segmentation for these clients, the firewall controls access to the endpoint device. An important case is IP-based wired payment card readers (for example, PCI compliance).

Unified policy enforcement: Converging both wired and wireless client traffic on the controller enables the network administrator to implement the same policies (user and traffic) for both wired and wireless clients. Unifying policies at the controller level has the benefit of providing deep visibility into the traffic, allowing administrators to consistently define and apply policies to the traffic, regardless of source (Figure 2-8).

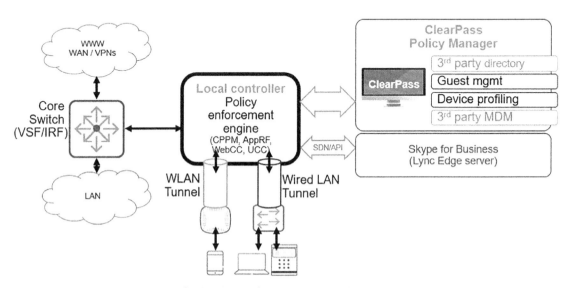

Figure 2-8 Unified policy enforcement using dynamic segmentation

Mobile First Access — Distributed reference design

The first two access designs had in common the presence of mobility controllers. This third design presents a fully distributed solution based on Aruba Instant APs (IAPs).

A group of IAPs connected to the same VLAN forms an Instant Cluster. Instant clusters implement a distributed mobility controller function with a controller node located inside each of the IAPs. Even when the feature richness of a mobility controller is not fully implemented on an instant cluster, these sets of IAPs can perform most of the functions of a controller including stateful application-layer firewall (Figure 2-9).

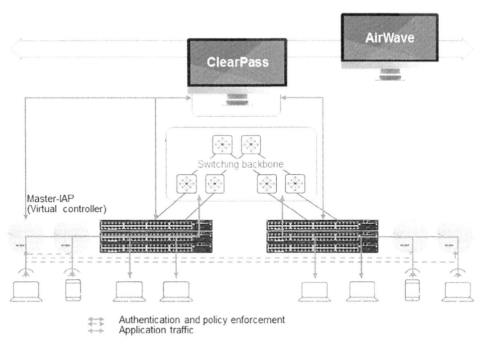

Figure 2-9 Distributed access reference design

The main characteristic of this design is its simplicity. All authentication, policy enforcement, and traffic forwarding are performed locally at the access layer.

Access components

Aruba Instant consists of a family of high-performance controller-less Instant Access Points (IAPs) that run the Aruba InstantOS to provide a distributed WLAN system.

In an Instant deployment, all IAPs on the same Layer 2 domain form a cluster with one dynamically elected AP that functions as the master. The master AP assumes the role of virtual controller (VC) within a cluster. The Graphical User Interface (GUI) to the VC provides local configuration and monitoring of an IAP cluster.

Aruba Instant is a distributed WLAN system with a completely distributed control and data plane. Each individual IAP handles the traffic for the clients that are associated with that IAP. Firewall policies and bandwidth control are also applied on a per-IAP basis. The flow of user traffic is not centralized to the VC. However, certain network functions, such as monitoring, firmware management, and source Network Address Translation (NAT) require a central entity within a cluster. The VC within a cluster functions as this central entity.

In an Aruba Instant cluster, if the master fails, another AP is elected as the master and assumes the role of VC. The master AP/VC failover time varies from 13 seconds to 100 seconds because the VC election algorithm also takes the CPU load on the IAPs in the network into account.

Instant clusters are limited to 128 APs and up to 2000 simultaneous clients. Because of this limit, the distributed campus access design can be applied to:

- Single building campus requiring less than 120 APs and with less than 1500 clients (reserving space for growth)
- Multibuilding campus in which a single cluster provides adequate coverage for any individual building and in which RF continuity between buildings is not critical.

Aruba AirWave (on-premise) or **Aruba Central** (off-premise/public cloud) can be used to provide centralized management of both the wired and wireless infrastructure in this design. One additional advantage of this design is that both the Aruba IAPs and switches support zero-touch provisioning (ZTP) using either AirWave or Central.

Edge optimization for Converged Wired + Wireless

Aruba offers an integrated access system that includes switches, APs, mobility controllers, AirWave or Central, and ClearPass. From the point of view of Aruba access switches, there is a set of features that optimize the edge for convergence of wired and wireless endpoints (Figure 2-10).

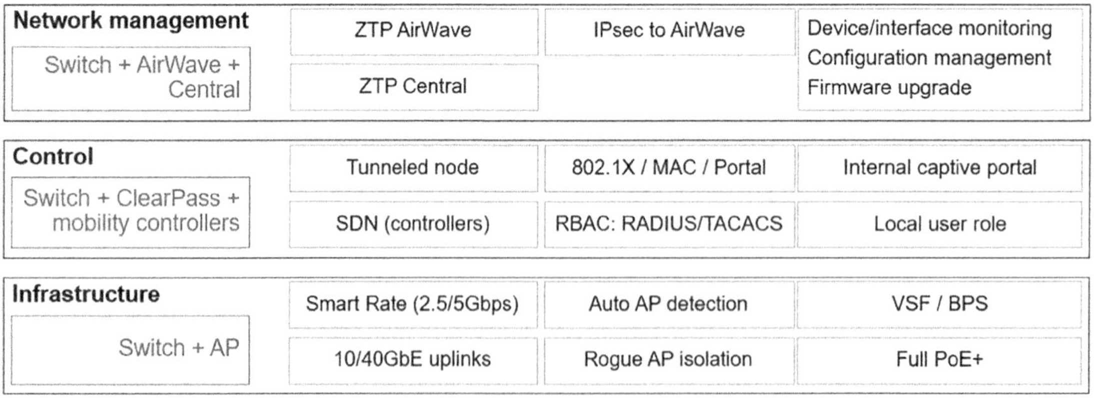

Network management			
Switch + AirWave + Central	ZTP AirWave	IPsec to AirWave	Device/interface monitoring Configuration management Firmware upgrade
	ZTP Central		

Control			
Switch + ClearPass + mobility controllers	Tunneled node	802.1X / MAC / Portal	Internal captive portal
	SDN (controllers)	RBAC: RADIUS/TACACS	Local user role

Infrastructure			
Switch + AP	Smart Rate (2.5/5Gbps)	Auto AP detection	VSF / BPS
	10/40GbE uplinks	Rogue AP isolation	Full PoE+

Figure 2-10 Aruba access switch integration features

AP integration

Smart Rate ports satisfies the bandwidth requirements of 802.11ac Wave 2 APs by providing a 2.5 Gbps link with PoE+. This aligns with the recently ratified NBASE-T standard (IEEE 802.3bz).

These 10/40 GbE uplinks satisfies a variety of uplink bandwidth requirements of access switches and stacks acting as AP aggregators, especially when Smart Rate ports are operating at 2.5 Gbps.

Auto AP detection simplifies the deployment of APs by automatically detecting Aruba APs and automatically applying a preconfigured profile to the corresponding port. The configuration profile includes a VLAN list (untagged and tagged), a CoS value, ingress and egress bandwidth limit utilization limit, PoE priority and maximum power, and speed and duplex mode (Figure 2-11).

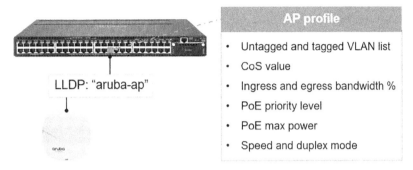

Figure 2-11 Auto AP detection

Rogue AP isolation (IAPs only) when deployed with IAPs, Aruba switches can isolate APs that do not belong to the cluster (Figure 2-12).

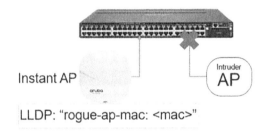

Figure 2-12 Rogue AP isolation with IAPs

Stacking—VSF/BPS even when stacking does not relate specifically to the AP integration and when paired with 10 GbE or 40 GbE uplinks, it optimizes the number of links between the access and the aggregation layer, reducing the number of cables and the number of ports on the aggregation switch.

Full PoE+ and redundant power depending on the APs model and quantity, and other PoE devices connected to the switch, the PoE budget plays an important role. And in many cases, redundant power can also be a requirement. The 5400R and 3810 Switch series both offer 30W per port and redundant power supplies.

ClearPass integration

Aruba switches integrate with ClearPass Policy Manager to provide a secure access environment, including RADIUS and TACACS device access control and 802.1X, MAC, and captive portal-based network access control. RADIUS CoA, local user roles, and captive portal are part of this integration.

ClearPass also plays an integral role in the execution of dynamic segmentation, as it allows wired endpoints to be discovered, profiled, and fingerprinted upon connection, before differentiated roles are dynamically delivered (including tunneled node, where appropriate) and any associated security policies enforced at the port-level of the wired edge network.

AirWave integration

AirWave offers the following Aruba switch management features:

- CPU, memory and interface monitoring
- Firmware upgrades
- Template-based configuration management including configuration auditing
- CDP/LLDP neighbors table
- CLI show commands

Zero-Touch Provisioning (ZTP)

Aruba switches can be automatically configured via the following methods:

DHCP + AirWave: By means of DHCP options, the switch receives the parameters necessary to contact AirWave: server IP address, secret key, device group, and folder. Using this information, the switch self-registers into AirWave and the template for that device groups/folder is applied, including any relevant variables.

Activate + AirWave: If the switch does not receive the options from DHCP, it contacts Activate. If the device has been registered in Activate, the switch receives the parameters necessary to contact AirWave: server IP address, secret key, device group, and folder. Using this information, the switch self-registers into AirWave and the template for that device groups/folder is applied, including any relevant variables.

Activate + Central: If the switch does not receive the options from DHCP, it contacts Activate. If the device has been registered in Activate via Central, the switch's connection is forwarded to Aruba Central from which it receives its configuration.

Mobile First Backbone—Collapsed and hierarchical reference designs

To connect the access system to the different network services and applications in the campus, network designers have to choose between a single-tier (collapsed) and a hierarchical backbone.

In a single-tier backbone, access switches are connected directly to a single-core switching node (Figure 2-13).

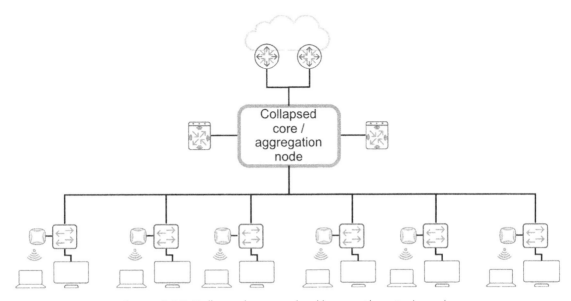

Figure 2-13 Collapsed campus backbone with a single node

In a hierarchical backbone, access switches are connected to aggregation switching nodes (also called distribution switching nodes), and the aggregation switching nodes are all connected to a core node (Figure 2-14).

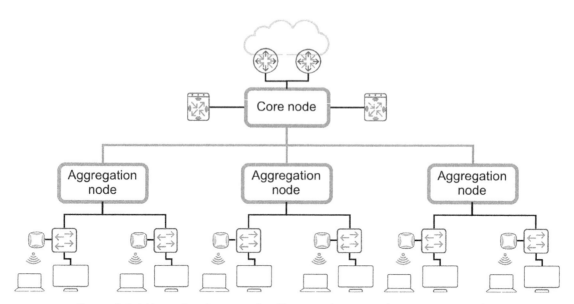

Figure 2-14 Hierarchical campus backbone with core and aggregation nodes

Historically, campus networks were designed using a hierarchical backbone for several reasons:

- To multiply the capacity of the different tables: MAC, ARP, routing, and others
- To multiply the Layer 3 forwarding capacity and improve performance
- To reduce the port density per node
- To simplify the cabling plant especially in multibuilding campuses

With the evolution of the switching technology, the first three items are no longer an issue. However, in large, multibuilding campuses, the cabling plant can be extremely complex if all access switches are connected directly to the core node, the cost of running multiple fiber optics cables between buildings may be higher than that of an aggregation switch.

Today, this is the main reason behind the choice of a hierarchical backbone.

Backbone components and virtual switching framework

Because the network backbone is critical, the best practice is to deploy backbone nodes in pairs of identical, interconnected devices (Figure 2-15).

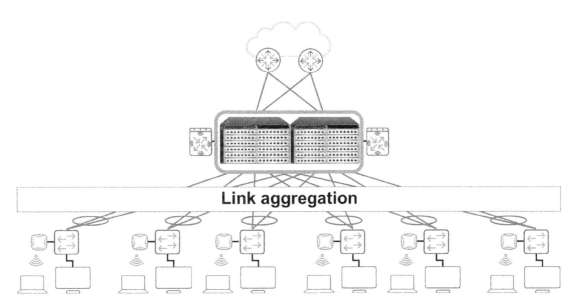

Figure 2-15 Typical campus backbone node

Aruba core and aggregation switches offer an approach of configuring both devices in the node as a single logical entity. In Aruba switches, this technology is called Virtual Switching Framework (VSF) (Figure 2-16).

Figure 2-16 VSF aggregates devices to create a single chassis-like switching framework

With VSF, a pair of backbone switches behaves as a single device, and there is no need to configure MSTP and VRRP or to configure routing in the access layer. Layer 2 redundancy is implemented by deploying link aggregation between neighbors, and Layer 3 redundancy is embedded in the VSF technology (Figure 2-17).

CHAPTER 2
Aruba Mobile First Campus Architecture

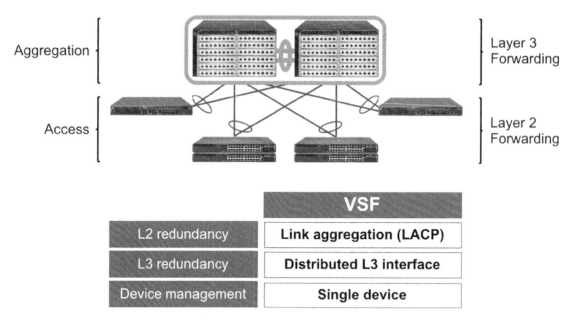

Figure 2-17 Access device connection to a VSF-based aggregation node

When the aggregation-switching node is a VSF fabric implementing Layer 2 switching in the access layer is the best practice. Figure 2-18 shows the different protocols implemented at each layer for unicast, multicast, and segmentation.

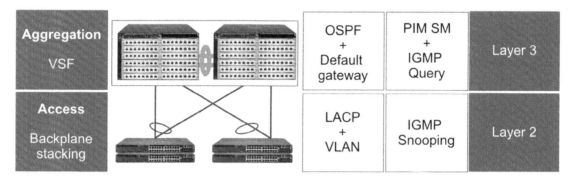

Figure 2-18 Layers 2 and 3 protocol mapping with VSF aggregation nodes

In some cases, network administrators or designers prefer to avoid creating a single control plane in the aggregation and core layers and choose the traditional dual-switch configuration. In that case, Layer 3 access is the best practice. Figure 2-19 shows the different protocols implemented at each layer for unicast, multicast, and segmentation.

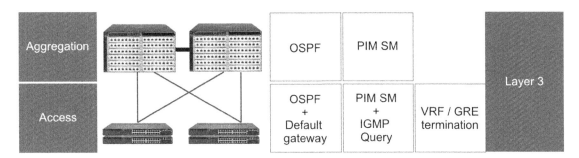

Figure 2-19 Protocol mapping with dual-switch aggregation nodes

Sample reference design — Single building campus

In a campus consisting of a single building, the network is a combination of one of the three access designs and a single core node. In this design the core-switching node connects directly to (Figure 2-20):

- All access switches (or stacks)
- Local mobility controllers (if a controller-based access solution is deployed)
- Campus edge routers
- Local server farm switches (if present)

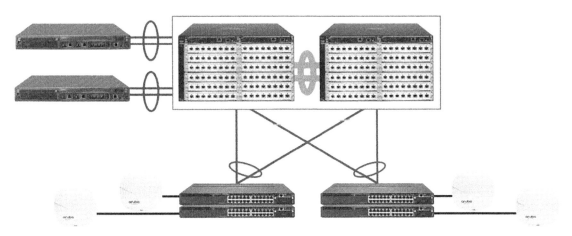

Figure 2-20 Single building campus reference design

Infrastructure components

Access system: Any of the three access reference designs described earlier can be used in a single-building campus. If managed APs and mobility controllers are being deployed (as opposed to IAPs), the recommendation is to connect controllers directly to the core.

Core (collapsed aggregation/core): Depending on the particular requirements, one of the switch series listed, in the previous section, can be deployed at the core.

Technologies and protocols

For redundancy reasons, the best practice is:

- To deploy two identical and interconnected core switches
- To connect all access switches with the same type of link to both core switches
- To connect each mobility controller/s to both core switches
- To connect each edge router/s to both core switches

Sample reference design—Multiple-building campus

A multibuilding campus consists of an access system, an aggregation switching system per building, and a central core switching system (Figure 2-21)

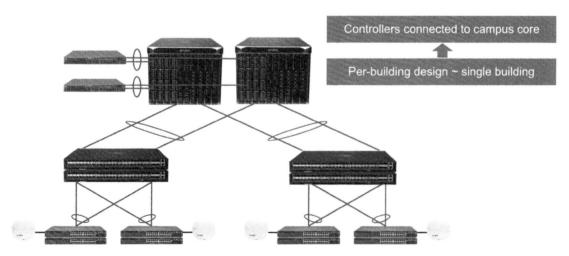

Figure 2-21 Multibuilding campus reference design

Infrastructure components

Access system: Any of the three access reference designs described earlier can be used in a single-building campus. If managed APs and mobility controller are being deployed (as opposed to IAPs), these controllers must be connected to the core.

Core and aggregation switching systems: The switches offered by Aruba for these two roles are the same described overhead for the single-building core.

Technologies and protocols

For redundancy reasons, the best practice is:

- To deploy two identical interconnected switches at both the core and aggregation layers
- To connect all access switches with the same type of link to both aggregation switches
- To connect all aggregation switches to both core switches
- To connect each mobility controller/s to both core switches
- To connect each edge router/s to both core switches

Single VLAN architecture for wireless clients

One important consideration when designing the core for a controller-based wireless network is that:

- Controllers are directly connected to the core
- Controllers are usually configured to forward in Layer 2
- The core switching system is the default gateway for all the wireless clients
- If access switches are also configured to forward in Layer 2, the core switching system is also the default gateway for all the wired clients
- As the number of clients grows, the size of the ARP table needs to be considered; if IPv6 is deployed, each device will require more than one entry in the ARP table (Figure 2-22)

CHAPTER 2
Aruba Mobile First Campus Architecture

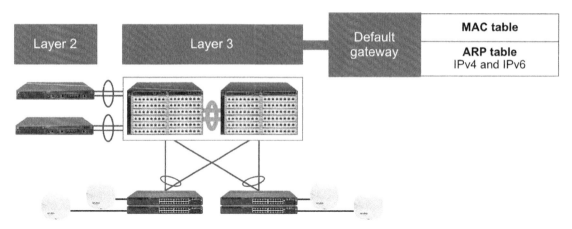

Figure 2-22 Core switch as default gateway for all wired and wireless clients

The increased use of IPv6 and a large number of mobile devices roaming across the campus has increased the complexity of WLAN design. Rather than using conventional design with VLAN pooling and multiple smaller subnets, Aruba believes a better way to design WLAN is to leverage a smaller number of larger subnets, or where appropriate, a single VLAN dedicated exclusively to transport wireless client traffic for a given SSID. For more information on best practices for adopting this particular design, refer to Single VLAN Architecture for WLAN—Validated Reference Design (VRD).

Summary

Aruba understands that rigid, static enterprise networks do not satisfy the needs of today's enterprises. Aruba's Mobile First network architecture for campus networks is, therefore, designed to allow people to move while staying connected and to enable enterprises to innovate without being tied to an outdated infrastructure. A campus architecture founded on integrated wired and wireless networks and best-of-breed solutions provides a scalable foundation to meet the growing expectation that "People move. Networks must follow."

References

Content updated and adapted from the following Aruba Design Reference Guide:

- **Aruba Mobile First Campus Reference Architecture**
    - https://community.arubanetworks.com/aruba/attachments/aruba/ForoenEspanol/2774/1/Aruba_Mobile-first_reference_architecture_guide.pdf

Additional detail on all solutions and technologies referenced in this chapter can be found at:

- **Aruba, a Hewlett Packard Enterprise Company**
    - http://www.arubanetworks.com

3 Software Defined Data Center (SDDC)

INTRODUCTION

The world is facing powerful major disruption, largely driven by new competitors with superior business models underpinned by the latest technology. Many firms are facing serious challenges in adapting quickly to new ways of creating and consuming IT services. They need to enable scalable business, fast entry into new markets, and agile and innovative services, as part of the strategic goals of a company. Some key drivers for digital transformation include the following:

- Business leaders wanting IT to be focused on business results, innovation, and continuous improvement.
- More effectual IT spending based on the need to meet business requirements.
- Organizations need to evolve to address the changing business landscapes.

Digital transformation can therefore be defined as the acceleration of business activities, processes, competencies, and models to fully leverage the changes and opportunities of digital technologies. Depending on where you are in your digital transformation journey, this chapter covers how HPE has simplified this transformation along two parallel streams, business and technology.

Digital Transformation in the Hybrid IT journey

451 Research reports regularly on *popular technologies and techniques* such as hybrid cloud, CICD, DevOps, containers and microservices, big data and analytics, mobility, and the Internet of Things (IoT) also known as Intelligent Edge. All of these are core enablers of hybrid IT and digital transformation.

Enterprises in pursuit of digital transformation will demand that their IT organization craft a holistic and uniform means for Lines of Business (LOBs) to deploy the right mix of portable workloads to the best execution venues in a hybrid IT architecture. As workloads shift to exploit the price/performance and elasticity, the current means to evaluate dependencies, integrate data and applications, orchestrate business processes, and enable adaptive collaboration will be challenged. APIs themselves are core integration enablers that now need to be strategically managed as assets with better tools to govern their design, quality, and use. All of these are somewhat overlooked and yet critical technologies for hybrid IT architecture. Moreover, any digital transformation initiative must be guided by a well-planned business strategy, that is, one that foretells the implications of change and can properly align hybrid IT resources with business needs.

CHAPTER 3
Software Defined Data Center (SDDC)

This is where the two parallel streams of **business** and **technology** transformation that we mentioned earlier come into play. The first one, business transformation, enables HPE Cloud Technology Partner (CTP) to assist global enterprise customers in transforming their traditional organizations to be agile digital structures encompassing people, processes, organization governance, security, and immersive DevOps-CICD. The second one, technical transformation, enables HPE to deliver API Driven integrated Hybrid IT solutions.

The Software-Defined Data Center (SDDC)

To enable technical transformation, a new framework is emerging supported by two core pillars:

- **API-driven technology** that supports traditional Service Oriented Architecture (SOA) long with microservices-based workloads for greater agility and speed
- **A DevOps model** of Continuous Integration and Continuous Delivery (CICD) for enhanced responsiveness to business requirements and demands

This newly evolved Software-Defined Data Center (SDDC) framework supports bi-modal delivery of both traditional SOA and microservices Container based workloads using an integrated architecture as shown in Figure 3-1.

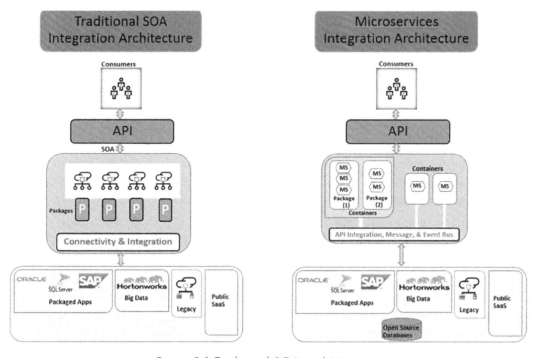

Figure 3-1 Traditional SOA and Microservices

The end result is a bimodal, Hybrid IT framework that can manage IT investments in a strategic and prioritized manner (Figure 3-1).

HPE OneSphere

Front-and-center of this integrated API Software-driven technology is **HPE OneSphere** that provides a hybrid CMP (Cloud Management Platform) using a Software Defined Data Center (SDDC) framework. OneSphere allows developers, IT Operations, and LOBs to expose agile services from internal IaaS, traditional-SOA environment, including VMware farms and clouds. This SDDC model is designed to drive agility, common/competitive integrated services, and lower the cost of complying with requirements from their LOBs. More details on API-driven SDC, SDS, and SDN/SDWAN especially around composability is provided later in the book. See the chapters describing HPE Synergy and Hyper Converged with SimpliVity.

Businesses need a true Hybrid IT platform, making it possible to develop and deploy workloads where they best fit based on business needs. HPE's strategy for OneSphere is to deliver a simplified turnkey platform for building, operating, and optimizing a Hybrid IT estate. This is a service that allows users to seamlessly compose and simply operate workloads. This strategy brings flexibility, choice, and cost-performance optimization by mixing and matching a range of infrastructure options.

Some key features and benefits of OneSphere include the following:

- Simplifies and accelerates the digital transformation from traditional to modern management to deploy apps
- Enables faster insights for IT Ops, developers, and LOB leaders
- Simplifies self-service provisioning
- Provides a consolidated application catalog service by integrating different catalog services
- Provides detailed analytics to track, categorize, and report on costs
- Gives the ability to view month-to-date and previous month resource cost information.

Enabling IT, LOB, and developers collaboration

One of the major challenges in digital transformation is to simplify IT, LOBs, and developers working together to accelerate this transformation as follows:

- **Improved speed and simplicity**—To respond to developers faster, IT Ops and administrators can deploy virtual machines, containers, and bare metal in minutes. As projects grow, HPE OneSphere enables rapid dynamic provisioning in a hybrid IT environment.
- **Accelerated application delivery**—With multi-tenant workspaces called **Projects**, HPE streamlines and speeds up application development and deployment with environments whereby

developers can, in a self-service process, provision, and access catalogs containing templates, services, and applications. A Project is also an abstract grouping of members or users that is used to control access to services and compute resources.

- **Greater cost efficiency**—HPE OneSphere provides a single view of usage and aggregate costs, so CIOs can control and optimize resources and spending. This solution also has the unique ability to provide "showback" costs by site, line of business, application or subscriber, and enable real-time self-service cost reporting to designated users.

Developers, IT Operations, and CIOs and business groups all benefit from HPE OneSphere. Some of the key HPE OneSphere features for developers and LOBs include:

- Unified experience across containers minimizing the need for specialized skills
- Built-in role-based collaboration Project workspaces designed for IT operations, business users and consumers of IT, such as developers, data scientists, and researchers
- Software-defined and API-driven virtual resource pool
- Subscription model
- On the private cloud side, OneSphere helps customers to define their own fixed cost model, which in turn can become the baseline for their internal chargeback

Accelerating digital transformation with bimodal IT

The research firm Gartner recently coined the term *bimodal IT* as a way to describe the need for IT organizations to support the diverse demands of the digital transformation that affects almost every enterprise. Many organizations are working on digital transformation in an effort to better engage with their customers and business partners and ultimately achieve a competitive advantage.

Gartner's bimodal IT model declares that there are two distinct ways for IT to operate in order to meet the demands of the digital economy. "Bimodal IT is the practice of managing two separate, coherent modes of IT delivery, one focused on stability and the other on agility. **Mode 1** is traditional and sequential, emphasizing safety and accuracy. **Mode 2** is exploratory and nonlinear, emphasizing agility and speed," according to the "Gartner Glossary."

DevOps is about adopting agile practices throughout the IT value chain, from business to development, up to IT operations supporting bimodal delivery workloads. For IT organizations that have already implemented agile methods to accelerate code delivery, DevOps is clearly the next logical step in their digital transformation. It is a path to deliver faster and more securely to the production environment. Eventually, the objective is to bring the final product to end-users as fast as possible. It is a new paradigm within IT that completely transforms the way of working between development and operations teams. The DevOps transformation challenge is about simultaneously managing the various dimensions of the transformation: cultural change, plan, build, and run processes, software delivery tooling, development and operations organizations, but also architecture and security.

The next sections are some of the examples of HPE's assets to simplify Mode1 and Mode2 leveraging DevOps to accelerate both traditional and microservices application development and deployment in their bimodal enterprise digital transformation.

For more detail on Gartner's bimodal view and Myths in accelerating Digital Transformation: https://www.gartner.com/smarterwithgartner/busting-bimodal-myths/

Example 1: SimpliVity microservices orchestration

Enterprises need to shift from a plan-driven to a value-driven management of the business outcomes, and that is the core of an Agile IT Organization design. Microservices and containers are major enablers for the following:

- Improving the scalability of software development and IT operation teams
- Increasing the ability to deliver faster and with better quality
- Improving the efficiency and management of IT services

The concepts of continuous integration and continuous delivery in a digital DevOps model is to effectively and efficiently design, create, and move into production software solutions: with the explicit notion that we should treat every change of code as a candidate release. Moreover, in recent years other concepts have emerged such as domain-driver design, on-demand virtualization, infrastructure automation, small autonomous team and systems-at-scale. microservices (small, autonomous services that work together) have emerged from these forces, as a trend or pattern, from real-world use. It is now understood that by using microservices architecture it is possible to deliver software much faster and embrace new technologies more easily.

Kubernetes has become the leading platform for powering modern microservices in recent years. Its popularity is driven by the many benefits it provides, including the following:

- Kubernetes encourages a modular and distributed architecture that increases the agility, availability, and scalability of the application
- Portability: Kubernetes works the same way, using the same images and configuration, no matter which data-center environment is being used.
- Open-source: Kubernetes is an open-source platform that developers can use without concerns of lock-in and is the most widely validated in the market today.

OneSphere Managed Kubernetes is provided as a Software as a Service (SaaS)-managed solution, with deployment, monitoring, troubleshooting, and upgrades performed by OneSphere, which also provides the operational Service Level Agreement (SLA) for Kubernetes management.

- A single view of multiple clusters
- Highly available, multi-master Kubernetes clusters that are automatically scaled-up and scaled-down based on workloads

- Common enterprise integration such as isolated namespaces; and the ability to deploy applications via Helm Charts that helps you define, upgrade, and install Kubernetes resources

Figure 3-2 is an example of accelerating DevOps and microservices using HPE SimpliVity. The hperconverged platform enables developers to deploy agile microservices and containers via OneSphere. The figure shows a bimodal architecture with multiple databases.

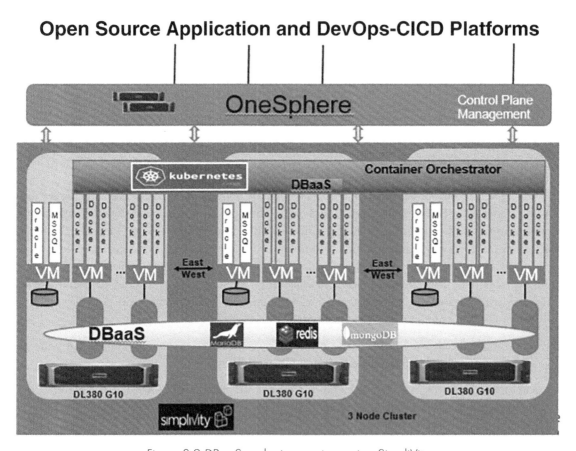

Figure 3-2 DBaaS and microservices using SimpliVity

Example 2: Synergy microservices and traditional composability

Figure 3-3 shows a different example of accelerating DevOps and microservices using HPE Synergy. In this example, developers are able to deploy agile microservices and containers via OneSphere in a composable bimodal architecture.

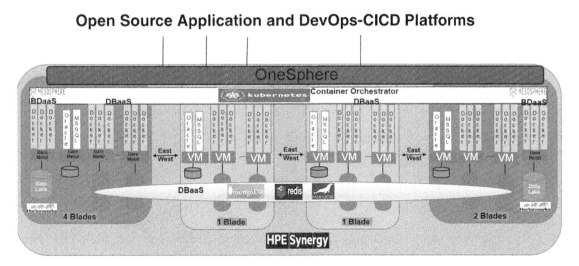

Figure 3-3 DBaaS and microservices using Synergy

Using OneSphere to build, operate, and manage hybrid IT

HPE OneSphere is a software-as-a-service (SaaS)-based management solution. It optionally can be bundled with HPE software-defined infrastructure solutions, including **HPE Synergy** composable infrastructure as well as **HPE SimpliVity** hyperconverged infrastructure. Through a unified view in HPE OneSphere, internal stakeholders such as IT operations, developers, and business executives can use hybrid IT.

OneSphere is built on a modern architecture using container-based services and Kubernetes orchestration. It is API-driven and can be integrated into an existing environment. The seed for OneSphere was planted with HPE's OneView, a management and automation facility for on-premises servers, storage, and networking. With OneSphere, HPE has extended that model to cover hybrid IT. The simplified deployment and management features of HPE OneSphere provide a cloud-like experience with on-premises infrastructure.

HPE consumption-based services

Capacity planning is difficult in many situations due to overprovisioning in IT to play it safe or underestimating future needs that leads to under provisioning. Both approaches require significant capital outlays and have long procurement cycles, limiting IT agility and restricting business operations. And it offloads much of the "undifferentiated heavy lifting" that comes with operating a data center. If your business uses on-premises IT for reasons such as security, control, and governance, you may hesitate to move to on-demand IT. But the good news is that HPE Pointnext can remove the burden of choice. HPE GreenLake Flex Capacity combines the simplicity, agility, and economics of hybrid IT with the security and performance benefits of on-premises IT. You determine your own "right mix" of Hybrid IT and workload placement without having to choose. With its agile pay-per-use service, HPE GreenLake Flex Capacity can help your IT organizations achieve the following:

- Avoid IT expenses stemming from overprovisioning
- Improve time to market by maintaining a safe buffer of capacity, ready for use when you need it
- Keep capacity ahead of demand with regular monitoring and a simple change order to replenish
- Pay for only the capacity used, not the capacity deployed
- Reduce IT risk with tailored support

HPE GreenLake solutions

HPE GreenLake provides modular, pre-packaged infrastructure choices to get you started faster and evolve your infrastructure ahead of your needs. These pre-packaged options make that process easy in that they cover a large range of needs, are pro-approved, and easy to quote and deliver. Once your current needs are met, the active capacity planning takes over, helping plan ahead of demand so that there is always a "buffer" of ready capacity in the data center, to handle growth. This cuts the cost of overprovisioning and eliminates the downtime and performance issues that result from lack of capacity.

HPE GreenLake is next generation of on-premises consumption-based services, going beyond pay-per-use infrastructure to pay-per-outcomes. HPE GreenLake is a suite of consumption-based solutions available for top workloads such as Big Data, backup, open database, SAP HANA®, and edge computing which deliver IT outcomes in a pay-per-use model in your data center or other location. Each solution comes with advisory and professional services options to further help you with solution design and integration into your environment.

Professional services

To accelerate time-to-value for HPE OneSphere, HPE Pointnext provides a high-level priority road map outlining key Hybrid IT business goals and next steps. Finally, a portfolio of options is available for enhanced call handling and expert support to provide the reliability, serviceability, and near-continuous availability. Figure 3-4 summarizes some of these services.

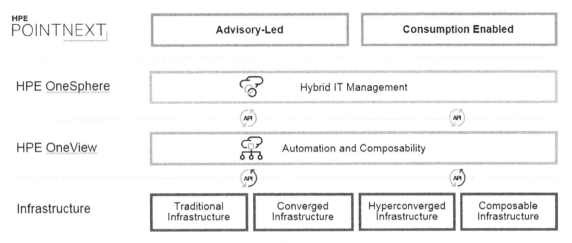

Figure 3-4 Sampling of key professional services

Summary

Transforming from a traditional environment to a digital enterprise organization often means your applications exist in multiple data centers, in a hybrid IT model, and at the edge of your network. Hybrid IT delivers many advantages, but it can also introduce new challenges in operational efficiency and multi-platform administration. A comprehensive, unified management solution helps remove these limitations to your successful digital transformation.

HPE enables enterprises to eliminate IT silos and reduce complexity with software-defined intelligence. You can simplify and automate your IT operations at a lower cost by transforming servers, storage, and networking into an API driven, software-defined infrastructure to address the followings key challenges:

- Optimize IT operations with continuous delivery
- Simplify operations and lower costs
- Develop and deploy apps faster in DevOps-CICD framework

To accelerate time-to-value using HPE OneSphere, HPE Pointnext can prepare your teams for success with expert professional and consumption-based services equipped to realize your business and technical goals for digital transformation.

4 HPE ProLiant Gen10 Advancements

INTRODUCTION

This chapter highlights the HPE Gen10 Server platform innovations. The topics covered include enhanced security, high-speed memory capacity with persistence, higher compute performance, increased in-server storage density, intelligent system tuning, and more efficient server management as described in the following list:

Enhanced security

- Protect, Detect, and Recover are the components of the enhanced Gen10 security covered in the upcoming section.

High-speed memory

- High-capacity data acceleration with flash-backed persistent memory.

Intelligent system tuning

- Performance tuning to enable more workloads on more cores at a given CPU frequency for greater application licensing efficiency.
- Predictable latency reduction and balanced workload optimization.

Higher levels of compute performance

- Next-generation industry standard CPUs with faster processing, higher speed memory access.
- Enhanced software-defined management and security.
- Enhanced GPU levels of performance and choice.

Increased in-server storage density

- Substantially greater Non-Volatile Memory express (NVMe) capacity for large write-intensive workloads needing advanced caching and/or tiers.

More efficient and easier server management

- Enables large-scale firmware deployment.
- Improved GUI to simplify management with industry standard Application Programming Interfaces (APIs).
- Easy system debug access.

CHAPTER 4
HPE ProLiant Gen10 Advancements

For details on how these enhancements dramatically improve IT security, systems performance, and operational efficiencies, refer to the appropriate sections.

Enhanced security

HPE is the first vendor to put silicon-based security into its industry standard servers, addressing firmware attacks, which are one of the biggest threats facing enterprises and governments today. Figure 4-1 illustrates a high-level depiction of iLO5 "silicon root of trust" process.

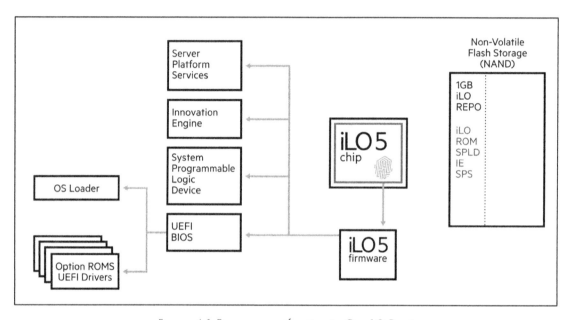

Figure 4-1 Firmware verification in Gen10 Boot

This groundbreaking silicon root of trust process is described in more detail beneath.

Protect, detect, and recover

The new integrated Lights Out 5th generation (iLO 5) remote management ASIC, embedded in most Gen10 servers, is manufactured with a hash code immutably burned into each iLO 5 chip. Coupled with digitally signed firmware images, this "digital fingerprint" embedded in the chip forms the foundation for a true silicon-based hardware root of trust.

The manufacture supply chain of the iLO 5 ASIC, as well as its firmware, is strictly controlled by HPE and is designed to enable multiple layers of verification:

1. When power is applied to a Gen10 server, the iLO 5 chip first verifies that the iLO 5 firmware image is authentic.

2. Assuming the image is validated, the iLO 5 firmware then verifies the authenticity of:

   a. The System ROM

   b. The System Programmable Logic Device (SPLD) firmware

   c. The Management Engine (ME) firmware

   d. Innovation Engine (IE) firmware

   If any of these firmware images are not properly authenticated, the firmware will not be executed.

3. Once these firmware components are verified, then the System ROM (UEFI BIOS) verifies the option ROMs and OS Boot Loader to extend the security "chain" up the software stack (Figure 4-2).

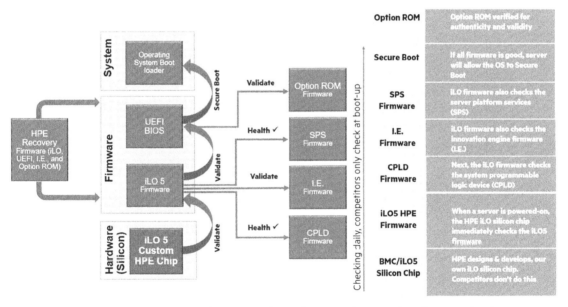

Figure 4-2 HPE Gen10 Silicon root of trust firmware boot and verification details

Administrator privileges

An administrator with the new "Recovery Set" user privilege can:

- Enable the iLO 5 to perform periodic firmware verification scans during runtime to ensure the continued integrity of the base firmware images.

- Schedule scans to run as frequently as once a day to once a year. These scans ensure that the base firmware images have not been corrupted or injected with malware while the server has been running.

- Select whether to:
  a. log the scan results when a problem is found during a scan, and/or
  b. log the scan results and automatically replace the invalid firmware with a "last known good" image.
- Maintain and update the firmware Recovery Set as needed.

 Note

The runtime firmware verification and automatic recovery features require a license for the new iLO Advanced Premium Security Edition. The firmware Recovery Set is initially installed at the HPE factory. It contains all five of the foundational images and is securely stored in the iLO Not AND (NAND).

Additional security options

A variety of additional security enhancements included in Gen10 are described beneath.

Security modes

Another new feature of iLO5 enables additional security modes (High, FIPS 140-2, CNSA). The "normal" mode of operation is now called Production Mode. The additional security modes include the following options:

- Any security mode higher than Production locks down specific host interfaces and mandates higher levels of cryptography in order to reduce attack surfaces.
- Any security mode higher than Production disables the ability for HPE System Update Manager (SUM) to directly install firmware remotely.
- Any security mode higher than High Security disables IPMI and SNMP v1 protocols.
- The Commercial NSA (CNSA) suite, also called SuiteB, has the highest level of cryptographic algorithms commercially available.

 Note

This features requires the new iLO Advanced Premium Security Edition.

Chassis intrusion protection device

A chassis intrusion detection device can be optionally purchased on most Gen10 server models. As long as the server is plugged into a power source, this device will detect if the chassis cover has been opened or closed, and log an event in the iLO5 integrated management log, even if the server is not powered on.

Trusted platform module

ProLiant Gen10 servers support an optional discrete Trusted Platform Module (TPM) that can be configured to operate in either a TPM version 1.2 or 2.0 mode. A no-cost (Factory Installed Option (FIO)) can be added to a configuration to instruct the factory to preset the desired TPM mode of operation. Intel's Platform Trust Technology (PTT,) which is a software implementation of TPM, is also supported, but is not certified.

Other security features

HPE supports the following security features and processes to ensure the secure configuration and operation of ProLiant Gen10 servers for organizations.

- HPE has introduced a new line of equipment racks that can accommodate a variety of locking methods, including third-party biometric lock options, for greater security.
- HPE is also extending security measures beyond the supply chain by selecting suppliers located in Trade Agreement Act (TAA) countries that are part of this program.
- A new deployment boot standard using HTTP/HTTPS is being driven by HPE and Intel to replace PXE boot, which many consider to have security vulnerabilities. This method still uses DHCP but is considerably more secure as it uses certificates for authentication.

For more security features and information please refer to the following URL: https://support.hpe.com/hpsc/doc/public/display?docId=a00018320en_us

Gen10 memory enhancements

At the time of writing, Gen10 servers offer the following enhancements to high-speed memory capacity with persistence.

Persistent memory NV-DIMM

Sixteen GigaByte (GB) non-volatile memory DIMM (NV-DIMM), enabling higher overall capacities, is supported on some Gen10 servers. NV-DIMMs combine Dynamic Random Access Memory (DRAM) and NAND flash on a single memory DIMM. During runtime, the NV-DIMMs DRAM region is accessed as normal memory. In the event of a power failure, the Smart Storage Battery provides the power necessary to offload the data in DRAM to the NAND flash region on the NV-DIMM. When power is restored, the data is moved back to its original location in DRAM. With increased capacity, users can safely operate on larger database tables and logs in-memory, resulting in greater performance and potentially lower per-core software licensing costs.

Scalable persistent memory

Scalable Persistent Memory (PMEM) is an HPE-exclusive technology that provides much greater capacity (~1 TB) than non-volatile DIMM's (NV-DIMMs). Initially supported in the DL380 Gen10 server, Scalable Persistent Memory technology combines elements of the BIOS, iLO5, DRAM, NVMe SSDs, and a new MicroUPS power supply. At initial setup, specific memory DIMMs are designated as persistent. Then, during normal operation, these DIMMs are accessed at DRAM speeds. The technical benefits include:

- The higher capacities made possible by Scalable Persistent Memory make it more feasible for users to load entire databases into memory for faster processing. This, in turn, provides an opportunity for users to reduce per-core software licensing costs if a server can accomplish the same, or more, work in the same, or less, amount of time with fewer processor cores.

- In the event of a power failure, the MicroUPS power supply provides the power necessary to offload the data in DRAM, designated as persistent, to a pair of NVMe drives that store the data until power is restored to the server. When power is restored, the data is moved back to its original location in DRAM.

- Scalable Persistent Memory has the advantage of operating at DRAM speeds and, because the NVMe drives are only written to during a power failure, the effective write endurance of these devices will be much greater than if they were being accessed as a standard block storage device.

Intelligent System Tuning (IST)

Gen10 servers possess several new Intelligent System Tuning capabilities described in this section. These provide the following technologies and features to increase system performance and reduce the time administrators spend tuning systems for specific workloads:

- **Jitter smoothing** and **core boosting** are HPE exclusives that increase performance levels.
- **Workload matching** simplifies the system administrator's task of tuning multiple BIOS parameters for specific workloads. The goal is to accomplish the same, or more, work using fewer processor cores. This reduces per-core software licensing costs.

 Note

IST features and capabilities require Intel Skylake processors and may vary depending on server model and processor.

Jitter smoothing

Intel's Turbo Boost technology incrementally increases the processor clock frequency above the base level in order to increase performance. The processor will run at these higher clock speeds as long as the processor can operate safely within its thermal limits. Intel calls this the Thermal Design Point (TDP.) The processor eventually needs to lower the clock speed to decrease heat dissipation and this continual process of raising and lowering the clock frequencies can occur many times per second. Generally, Turbo Boost will improve performance, although this increase in performance is opportunistic and can be unpredictable. Each time the processor changes clock speed up or down, it must suspend executing instructions. This, in turn, results in short periods of time, approximately 10–20 microseconds each, during which the processor is idle. This is referred to as "jitter" or "white space." If these frequency shifts occur too often Turbo Boost can actually negatively impact overall performance by having too many periods of jitter.

Many applications, such as high-frequency trading applications, require deterministic, predictable performance and cannot tolerate this execution jitter. As a result, Turbo Boost is generally disabled for these types of workloads. However, with jitter smoothing technology in select Gen10 servers, administrators can reduce or eliminate workload jitter. When enabled and configured for autotuned mode, jitter smoothing works by automatically adjusting the clock frequency during runtime to the highest possible speed that allows the processor to operate safely and not make further frequency changes. A manual-tune mode is also available that allows administrators to specify a frequency above the base and manually tune for jitter reduction.

For many workloads, jitter smoothing, in combination with Turbo Boost, can offer higher and predictable performance at a clock speed higher than the base frequency.

Core boosting

HPE has worked closely with Intel to gain exclusive access to registers on select Skylake processors for greater performance when Turbo Boost is enabled. HPE's patented core boost technology can use more active cores at higher turbo frequencies, without over-clocking, than the standard Turbo Boost algorithms. By not over-clocking, Intel's reliability levels and warranties are maintained. Intel obviously sells processor chips to many server manufacturers, and they rely on the respective design engineers for those manufacturers to provide a server chassis environment that allows the processor to operate within its TDP. By monitoring and tightly controlling the thermals of the processor as well as the server chassis, HPE core boost is able to maintain more active cores at higher clock frequencies providing greater performance than Intel's "straight-line" approach.

By using HPE core boost technology, users can achieve higher levels of performance.

Workload matching

Gen10 servers have several preconfigured workload profiles from which an administrator can select to automatically configure relevant BIOS settings in order to optimize the specified workload. This greatly reduces the time and potential guesswork administrators need to research and adjust individual BIOS settings. For example, administrators can select a profile for low latency compute, high I/O throughput, or power-efficient virtualization, among others.

HPE has published a white paper describing these workload profiles along with the BIOS settings that are affected by a given profile, providing guidance to administrators as shown at the following URL: https://support.hpe.com/hpsc/doc/public/display?docId=a00018313en_us

Higher levels of compute performance

Gen10 includes many compute options that provide increased performance and some unique capability with graphics processors as described in this section.

Enhanced processors

Inherent to this new Skylake microarchitecture are performance enhancements. Intel has changed the memory subsystem from four memory channels per processor to six as well as increasing the maximum transfer speed from 2400 Memory Transfers per second (MT/s) to 2666 MT/s. Although the total number of memory slots remains the same from the previous generation, increasing the number of channels can provide increased overall bandwidth. The BIOS of a system can play a significant role in its overall performance.

HPE engineers write and optimize each platform's BIOS image and, as a result, best-in-class benchmark results have already been achieved by ProLiant Gen10 servers as shown at the following URL: (http://h17007.www1.hpe.com/us/en/enterprise/servers/benchmarks/index.aspx#.Wbq65U3ru1s)

Beyond the processor and memory complex, each Skylake processor also has eight more PCIe 3.0 lanes than the previous generation, increasing overall I/O performance.

AMD has announced a reentry into the x86/x64 server processor market. The new EPYC processors have up to 32 cores 64 threads with eight memory channels supporting 16 memory slots per processor, with up to two processors per system. This new architecture does have a unique security feature with the ability to encrypt individual Virtual Machines (VMs) in memory. Coupled with AMD's Secure Memory Encryption (SME) technology, and Secure Encryption Virtualization (SEV), a security processor attached to the memory controller can assign a unique cryptographic key to individual VMs thereby isolating them from the hypervisor and other VMs.

Support for Graphical Processing Units

An increasing number of modern applications are taking advantage of Graphical Processing Units (GPUs) for greater compute power. Emerging technologies, such as Artificial Intelligence (AI) and Deep Learning applications often require the processing power of GPUs. More traditional workloads, including seismic analysis, weather prediction, and 2D/3D high-definition visualization also benefit from more powerful GPUs. In recent years, NVIDIA and other manufacturers have enabled these devices to be virtualized and shared among multiple remote desktop users whose workloads require or benefit from graphics acceleration, lowering the cost per desktop as a result.

In several Gen10 server platforms, HPE is enabling support for a greater number of GPUs. For example, the DL380 Gen10 can support up to three double wide or five single wide GPUs, versus two and three, respectively, in the previous generation, thereby increasing computational capabilities beyond the general purpose server processors supplied by Intel and AMD. Because these GPUs frequently draw more power and generate more heat than the main processors, HPE engineers perform extensive thermal testing during product design to ensure proper cooling for maximum reliability.

Increased in-server storage density

In order to increase the space available for storage devices, design engineers reduced the physical size of Gen10 FlexSlot power supplies by approximately 25% without sacrificing wattage capacity. As a result, a wider variety and quantity of internal and rear-mounted storage devices are now available in many platforms.

Other efficiencies were gained by shrinking the storage devices themselves. One example is the 340GB SATA Dual Micro Form Factor (uFF) M.2 SSD device. It is comprised of two independent hot-swappable SSD devices within a single SFF drive carrier. This enables a half-height Gen10 blade, for example, with two internal drive bays to now have RAID5 capability using four independent hot-swappable internal SSDs.

In many cases, Gen10 also has increased support for NVMe devices in terms of the number of devices as well as capacity while also lowering latency by eliminating PCIe switching.

All of this increased storage density is enhanced with Gen10 Smart Array controllers. Internal lab testing performed in January 2017 showed that Gen10 controllers have approximately 65% greater performance when compared to previous generation of controllers. They also consumed approximately 45% less power while providing additional functionality. The Gen10 Smart Array controllers can operate in RAID and HBA modes simultaneously. The new controllers carry forward the ability to encrypt data at rest, including data in the write cache, and configure one or more SSD's as SmartCache for rotating media.

More efficient and easier server management

Many of the enhancements to Gen10 have been made easier to manage by the new capabilities for server management that are described in this section.

ASIC enhancements

In addition to the new security and performance tuning features, the iLO5 Application-Specific Integrated Circuit (ASIC) provides the following set of enhancements:

- A faster processor and more memory so that the remote console and virtual media functions perform better.

- Improvements have been made so that a Gen10 server takes approximately 66% less time to Power on Self-Test (POST).

- The iLO5 ASIC does not rely on operating system-based management agents and has visibility to more devices than previous generations.

- A new iLO Front Service Port on Gen10 DL rackmount servers allows support personnel to connect a laptop to the iLO and eliminate the need for a KVM crash cart. If enabled, this port could also be used to offload Active Health System logs or update iLO firmware.

- Enhanced integration between iLO5 and embedded Intelligent Provisioning has led to access of all BIOS settings from within Intelligent Provisioning as well as faster launch times.

- Although not exclusive to iLO5 and Gen10 servers, the recently introduced iLO Amplifier Pack can increase the efficiency of administrators who manage hundreds or thousands of Gen8 (iLO4) or newer servers.

Unified CLI

The UEFI BIOS in HPE Gen10 servers now has a unified CLI. The RESTful API schema has moved completely to the Distributed Management Task Force (DMTF) Redfish standard (https://www.dmtf.org/standards/redfish). The Gen10 Smart Array storage controllers can now be configured in the UEFI BIOS either manually, or via RESTful API (an OEM extension of the Redfish standard). This is in addition to the traditional Smart Storage Administrator tool.

Infrastructure management with HPE OneView

A new version of OneView, HPE's infrastructure management application, was released which supports not only the new Gen10 platforms, but some additional Gen9 server models and storage products as well.

Example of Gen10 workloads enhancements

Gen10 can be used for almost any workload. In this section, we show the Gen10 configuration used to optimize a sample workload on Microsoft SQL Server using Persistent Memory.

Solution overview

Persistent Memory NVDIMMs in a SQL Server workload can reduce the database storage bottlenecks by increasing speed up to 4x on Input/output (I/O) as well as deliver better processor utilization with byte addressable storage. The following configuration would enhance performance of SQL tail-log backup. Tail-log backups are the capture of any records that have not been part of a backup to prevent the loss of those records.

Server configuration

- 1 x HPE DL380 with two Intel Skylake processors
- 128 GB Memory
- 1 x 8 GB NVDIMM will be used to store the tail of the log
- 2 x 400 GB SATA SSD will be used as the store for the database files
- 1 x 400 GB NVMe SSD will be used to store both logs
- Software Configuration – Figure 4-3
- Windows Server 2016 TP5
- Microsoft SQL Server 2016 RC3

Configured with the following parameters:

- SQL Tables are stored on the two 400 GB SATA SSDs and configured as RAID0 (Striped)
- SQL tail-log is enabled
- Table Size is configured to match data and log storage capacities
- Threads: 1 per windows logical processor
- SQL queries: Create, Insert, Update
- SQL PerfCollector: None
- Batch Size: 1
- Row Size: 32 GB

Figure 4-3 shows the results of using NVDIMM versus SSD for tail-log.

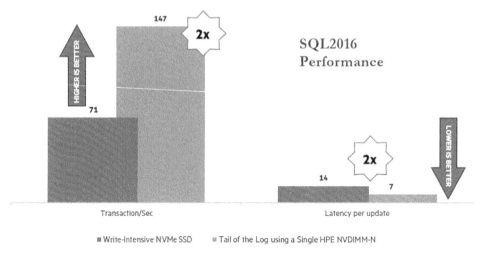

Figure 4-3 Tail-Log Performance of SSD versus NVDIMM

The addition of a single NVDIMM has improved the transaction speed by 2x as well as lowered the latency by 2x.

The modest addition of one NVDIMM has helped deliver a superior Microsoft SQL 2016 workload performance, accelerate the application response time, and more importantly, the end-user experience.

Database checkpoint and restore

The introduction of HPE Persistent Memory NVDIMMs to a database checkpoint-restore has shown a significant performance improvement as shown in the following two figures:

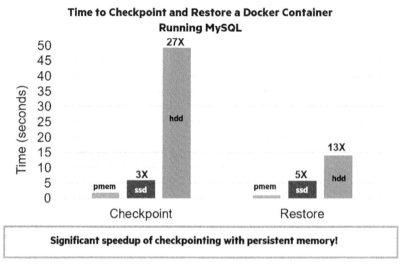

Figure 4-4 Database checkpoint and restore performance in MySQL

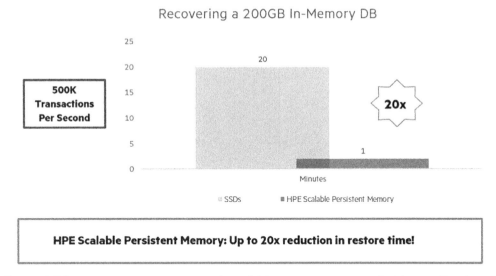

Figure 4-5 Restore time improvement with scalable persistent memory (In-Memory Database)

Both of these diagrams show performance increases with persistent memory. Figure 4-4 depicts Checkpoint and Restore performance and Figure 4-5 shows Recovery time improvement. Figure 4-4 includes performance data for persistent memory, SSD, and HDD. Figure 4-5 shows the performance increase with scalable persistent memory.

Summary

This chapter highlighted many innovations in HPE Gen10 Server platforms that were selected only after extensive research and input from customers. These innovations cover a variety of technical areas that are designed to target important industry focus areas such as security (everyone is interested in enhanced security), increased performance (everyone wants and needs a faster server), nonvolatile memory (this can change the way that you run mission-critical applications), and so on. Thanks to Gen10 servers, it is now possible to chart a less costly course to positively impact the way that critical applications are run and, at the same time, transform the way in which servers are protected.

5 The Power of Azure in your Data Center: HPE ProLiant for Microsoft Azure Stack

INTRODUCTION

In this chapter, we will cover the HPE ProLiant for Microsoft Azure Stack solution that enables organizations to deploy an Azure hybrid cloud platform to an on-premises data center. This hybrid IT solution combines the benefits of Azure public cloud, such as agility and scalability, with the benefits of on-premises data center technologies such as control, performance, and security.

Microsoft's Hybrid Cloud Platform

Identifying the right mix of public and private cloud deployment is a key initiative in almost every organization. The right mix results in a hybrid IT strategy that ideally suits the unique requirements of each organization. At the heart of many such activities is Microsoft Azure public cloud that since 2008 has enabled countless businesses to move faster and achieve more. When it comes to an overall hybrid IT strategy, however, there are many more considerations involved beyond simply speed and performance. These include but are not limited to regulatory compliance, data sovereignty, security, cost, latency, and many other issues, both business and technical. In many cases, these considerations hinder cloud adoption. To address these considerations and challenges, Microsoft introduced Azure Stack, bringing Azure's ecosystem and services to the data center. Figure 5-1 depicts Microsoft's Hybrid Cloud platform.

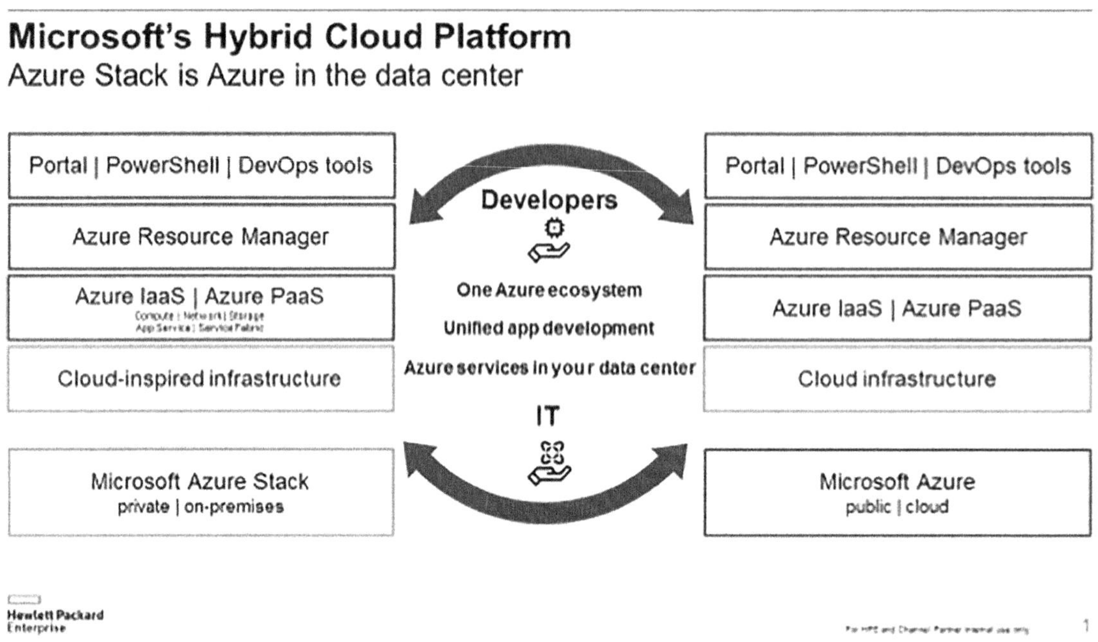

CHAPTER 5
The Power of Azure in your Data Center: HPE ProLiant for Microsoft Azure Stack

Microsoft's Hybrid Cloud Platform
Azure Stack is Azure in the data center

Private / On-premises	Shared	Public / Cloud
Portal | PowerShell | DevOps tools	Developers	Portal | PowerShell | DevOps tools
Azure Resource Manager	One Azure ecosystem	Azure Resource Manager
Azure IaaS | Azure PaaS (Compute | Network | Storage, App Service | Service Fabric)	Unified app development	Azure IaaS | Azure PaaS
Cloud-inspired infrastructure	Azure services in your data center	Cloud infrastructure
Microsoft Azure Stack — private | on-premises	IT	Microsoft Azure — public | cloud

Hewlett Packard Enterprise

Figure 5-1 Microsoft Hybrid Cloud Platform

As you can see in Figure 5-1, there is a great deal of synergy between the public cloud environment (shown on the right of the figure) and private cloud environment (shown on the left).

Solution overview

Co-engineered by Hewlett Packard Enterprise and Microsoft, the HPE ProLiant for Microsoft Azure Stack solution provides organizations with an Azure hybrid cloud platform in an on-premises data center that is fully compatible with Azure Public Cloud services. This hybrid cloud solution delivers the speed, agility, and simplicity of a public cloud, combined with the cost-effectiveness and security of a powerful on-premises private cloud. Adding to the benefits is consistency with Microsoft Azure public cloud services.

HPE ProLiant for Microsoft Azure Stack is an integrated system that delivers Azure-compatible, software-defined infrastructure as a Service (IaaS) and Platform as a Service (PaaS) on HPE hardware. This means you can transform on-premises data center resources in a hybrid cloud environment or build your own on-premises private cloud service, which provides more agility to leverage applications designed and built for the cloud.

Users can quickly provision and scale services with the same self-service experience as Azure. For example, Microsoft Azure Stack accelerates DevOps by providing a rich Azure ecosystem of resources. Application developers can maximize their productivity using a write-once, deploy to Azure or Azure Stack approach. Using APIs that are identical between the public and private Microsoft Azure offerings. You can create applications based on your choices of open source or .NET technology.

Hardware inventory

In order to deploy an on-premises solution, the following tested high-level hardware components were selected to run on Azure Stack with the following characteristics:

- Two HPE Networking 5900 series top-of-rack switches
- One HPE Networking 5900 series management switch
- One HPE ProLiant DL360 Gen 9 hardware lifecycle host
- 4–12 configurable HPE ProLiant DL380 Gen 9 compute nodes
- Multiple racking options
- Multiple power options
- Optional KVM and LCD Console
- Factory integration
- On-site deployment
- Multiple options for Azure Stack and Azure Stack software support
- Multiple options for infrastructure support

Figure 5-2 depicts the main physical components in a single rack diagram.

CHAPTER 5
The Power of Azure in your Data Center: HPE ProLiant for Microsoft Azure Stack

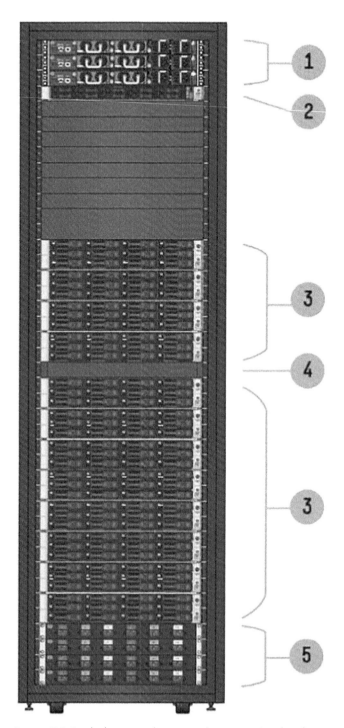

Figure 5-2 Rack diagram depicting the Azure Stack solution

The key below lists out each of the numbered components:

1. Solution switches (3). Defaults: HPE Ethernet Switch 5900AF 48XG 4QSFP+ (2) and HPE Ethernet Switch 5900AF-48 G -4XG-2QSFP+ (1)
2. HPE ProLiant DL360 Gen9 with Microsoft Azure Stack hardware lifecycle host (1)
3. HPE ProLiant DL380 Gen9 with Microsoft Azure Stack Nodes (4–12)
4. Optional: 8 or 16 port KVM switch and optional LCD console (1)
5. Power Distribution Units (2 or 4)

Key use cases

As mentioned previously, Microsoft has designed Azure Stack to address the challenges of hybrid IT. The following section is a sampling of some key requirements and challenges and the way in which Azure Stack addresses them.

Ensure compliance, data sovereignty, and security

While many organizations enjoy the ease, flexibility, and cost model of a public cloud, they are hesitant when it comes to using it for specific applications and data. Azure Stack is designed to address those concerns. It provides easy access to services that must meet specific compliance, data sovereignty, and security requirements. Going back to our original example in the introduction of this chapter, Europe has data sovereignty regulations that require data to be kept either within the European Union or within a country's borders. Azure Stack enables you to run the same service across multiple countries, as you would using a public cloud, but meet data sovereignty requirements by deploying the same application in a data center located in each country, thereby ensuring personal data is kept within that country's borders.

Maximize performance

If your applications, such as analytics, demand high levels of performance, public cloud may not offer the results you are looking for. For example, using public cloud to analyze large sets of data located in your data center requires that you first upload the data to the public cloud before analysis can take place. But the upload time may be prohibitive and not return the results as quickly as you need to meet requirements. Public cloud may also not meet your expectations if your applications require low or consistent latency. In both cases, running your workload in an Azure Stack on premises may meet your requirements.

Connect edge and disconnected applications

In some cases, you may be running applications either on the edge or ones that disconnect from your datacenter for a period of time. Some examples of applications that disconnect for a period of time are those that run on equipment in transit, such as cruise ships or freight trains, in a mini datacenter to manage the operations of the remotely. When the equipment is in transit, the mini data center on the equipment is connected to the main datacenter. But when the equipment is in transit, the mini datacenter runs disconnected from the main datacenter. Rather than having to wait to analyze data when the equipment returns to a central location, running Azure Stack enables equipment in transit to perform local, on-board analysis of the data. Results can then be uploaded to the main datacenter upon returning to the central location.

Accelerate modern application development

Modern, cloud-native applications are developed to run as micro-services in many different environments. Developers do not want to have to use a different set of development tools for different micro-services just because they run in a different environment. The fastest and most efficient option is to develop applications using a consistent set of tools and then deploy the application to wherever it is required. Azure Stack and Azure are API-compatible, eliminating that concern. A common API allows developers to develop applications once, and then easily deploy them to either Azure public cloud or Azure Stack running in a private cloud, with no changes to the application.

Summary

For more than 30 years, HPE and Microsoft have been helping their joint customers optimize their IT environments and leverage new consumption models to accelerate their desired business outcomes. An outstanding example of this is HPE ProLiant for Microsoft Azure Stack, an integrated system that brings the power of Azure's cloud to the data center. This hybrid IT solution provides many ground-breaking benefits, including:

- **Unmatched flexibility and configurability**

    With flexible configuration options, HPE ProLiant for Microsoft Azure Stack will fit seamlessly into any existing environment. The various options include:

    - The processor type that is right for the workload
    - Choice of memory
    - Scalable storage capacity
    - Support for third-party networking switches, power supplies, and rack options

- **Exceptional speed and performance**

    HPE ProLiant for Microsoft Azure Stack is uniquely architected to achieve both high capacity at 768 GB RAM, and high performance at full 2400 MHz memory speed, increasing memory bandwidth by up to 28% compared to other, same capacity solutions. This enables you to run more workloads—even faster.

- **Consumption-based IT for Azure Stack**

    HPE GreenLake Flex Capacity enables pay-per-use pricing reducing costs by leveraging cloud-style economics using a consumption-based model. HPE GreenLake Flex Capacity gives you the cloud you need with the following features and benefits:

    - Rapid scalability
    - Variable costs aligned to metered usage
    - No upfront expenses
    - Enterprise-grade support
    - One monthly bill

- **Professional services and expertise**

    The HPE ProLiant for Microsoft Azure Stack solution enables you to access the collective expertise of over 4000 HPE experts trained on Azure and hybrid cloud to answer any questions you have and give you the support you need. Take advantage of the experts to:

    - Develop the best hybrid cloud strategy for your company.
    - Deliver professional services to meet your use case, design, and implementation needs
    - Deploy your solution with confidence

6 HPE SimpliVity Hyperconverged Infrastructure

INTRODUCTION

Rapid proliferation of applications and the increasing cost of maintaining legacy infrastructure causes significant IT challenges for many organizations. With HPE SimpliVity, you can streamline and enable IT operations at a fraction of the cost of traditional and public cloud solutions by implementing advanced data services using a single, integrated all-flash solution. This simple and efficient hyperconverged platform reduces the overall costs and complexity of your infrastructure while delivering the advanced IT capabilities modern businesses require to stay competitive.

Solution overview

The HPE SimpliVity hyperconverged infrastructure employs the HPE SimpliVity Data Virtualization Platform (DVP) to provide scale-out enterprise capabilities to core datacenter workloads. The DVP converges the compute, storage, and data protection into an appliance that provides the following features and benefits:

- **Efficient data storage** through deduplication, compression, and optimization.
- **Predictable performance** by eliminating duplicate storage Input/Output (IO) and offloading data service operations to the HPE SimpliVity Accelerator Card that is installed in each SimpliVity appliance.
- **Native Data Protection and Data Recovery.** Data protection is policy-based and VM-centric and can be configured to meet Recovery Point Objectives (RPO), Recovery Time Objectives (RTO), data retention requirements, and efficient offsite data replication.
- **Simplified management** integrated into hypervisor management tools. All provisioning, data protection policy management, data recovery operations, and virtual machine workload management are accomplished through a single interface.

Solution components

An HPE SimpliVity hyperconverged infrastructure is comprised of the following components.

HPE SimpliVity appliance

The HPE SimpliVity appliance is also commonly referred to as a node or a host. It is built on the HPE DL380 Gen10 server and can be configured to meet CPU, memory, networking, storage performance, and storage capacity requirements.

Each HPE SimpliVity appliance includes the SimpliVity Accelerator Card that is a purpose-built PCIe card with a Field Programmable Gate Array (FPGA) processor, DRAM, flash storage, and super capacitors. The SimpliVity Accelerator Card is passed directly to the SimpliVity Virtual Controller, a virtual machine that runs on each appliance. The FPGA and DRAM provide hardware acceleration of the DVP data services, while the super capacitors provide power to allow for data to be de-staged from DRAM to persistent flash if there is an unplanned power loss.

Hypervisor cluster

HPE SimpliVity appliances appear as, and are managed as, hypervisor hosts. Hosts are organized in clusters. Multiple SimpliVity appliances configured in a cluster provide high-availability data stored within the cluster. This cluster configuration aligns with hypervisor clustering for both compute availability and compute resource distribution. A cluster is scaled by adding SimpliVity appliances to the cluster.

HPE SimpliVity federation

The HPE SimpliVity federation is the management and data mobility boundary of an HPE SimpliVity infrastructure. The federation can be made up of a single or multiple clusters of HPE SimpliVity appliances. The federation is managed using the vSphere Web Client. Multiple vCenter Servers can be deployed in Enhanced Linked Mode to support multiple management domains across the federation.

The image in Figure 6-1 displays an example HPE SimpliVity Federation design with multiple clusters across multiple sites to provide resources for production workloads, development and test workloads, and offsite disaster recovery.

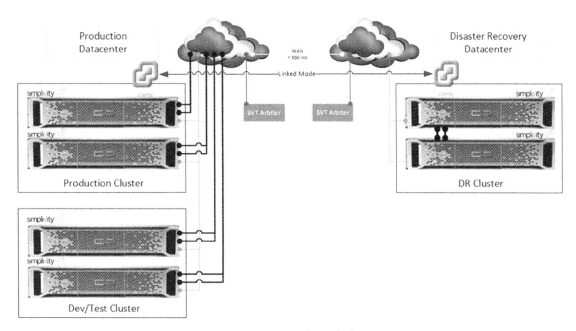

Figure 6-1 HPE SimpliVity federation

Architectural features and benefits

The upcoming sections cover the features and benefits of the HPE SimpliVity solution.

Storage efficiency

The Data Virtualization Platform (DVP) is the data architecture that powers the HPE SimpliVity hyperconverged infrastructure. It provides inline deduplication, compression, and optimization, of all data at inception, globally. DVP's efficiencies deliver the following benefits:

- Near-instant cloning of virtual machines using HPE SimpliVity Rapid Clone or VMware cloning.
- Local and remote protection of virtual machines with policy-based virtual machine backups.
- WAN efficient replication between HPE SimpliVity clusters.
- Quick recovery of virtual machine images from local or remote backups.

Inline deduplication and compression allow the storage capacity of the HPE SimpliVity appliance to exceed the physical usable capacity of the appliance. Table 6-1 outlines the physical storage configuration of the currently available HPE SimpliVity appliances.

CHAPTER 6
HPE SimpliVity Hyperconverged Infrastructure

Table 6-1 HPE SimpliVity appliance storage configuration and sizing

SimpliVity appliance	Drives: Number/Size/Type	RAID	Usable capacity*
Extra Small	5 × 900 GB SSD	RAID5	2.9 TB
Small	5 × 1.92 TB SSD	RAID5	6.0 TB
Medium	9 × 1.92 TB SSD	RAID6	10.9 TB
Large	12 × 1.92 TB SSD	RAID6	15.8 TB

* Usable capacity after RAID overhead and overhead associated with formatting.

The efficiency of the DVP provides dense capacity storage and IO avoidance. This results in a much smaller capacity footprint along with performance improvements. During the initial ingestion of data of general-purpose virtual workloads, we can assume a deduplication efficiency ratio of 1.5:1 and a compression ratio of 1.5:1. Combined this equals an overall efficiency of 2.25:1. This results in 2.25 TB of data, consuming 1 TB physical storage. The deduplication and compression efficiency ratio can vary dependent on a customer's data type, for example, images and videos may not compress or deduplicate well and adjustments will need to be made to the ratio if data of these types are present in the environment.

To calculate the total effective capacity of a SimpliVity cluster, we use the number of nodes in the cluster (N), the usable capacity of single node in the cluster (U), and the combined deduplication and compression efficiency (E). The capacity is then divided by 2 for HA. The complete formula for calculating the effective capacity of an HPE SimpliVity cluster is:

$(N \times U \times E)/2$ = Effective Capacity

For example, the effective capacity of an HPE SimpliVity Cluster with 4 × Medium appliances with a combined efficiency ratio of 2.25 would be calculated using $(4 \times 10.9 \times 2.25)/2$ that equals approximately 49 TB.

The logical capacity of the environment will be much higher than the effective capacity when HPE SimpliVity backups or clones are used. HPE SimpliVity backups and clones consume no more space than source at the point in time the backup or clone is taken.

The image in Figure 6-2 is a screen capture that displays the data efficiency of a current HPE SimpliVity customer.

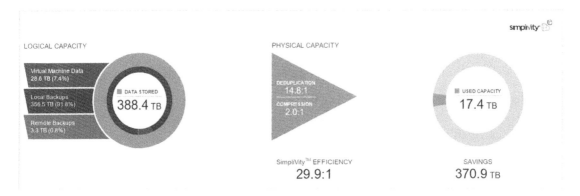

Figure 6-2 HPE SimpliVity data efficiency

The customer is storing ~388 TB of data in a 17.4 TB physical footprint. The capacity savings is nearly 371 TB, and this is also 371 TB that never had to be written to physical disk—all of this IO was completely avoided.

Configurable compute resources

Compute resources can be configured to meet the requirements of virtualized workloads. Determining compute configurations for the HPE SimpliVity appliance is similar to designing for traditional virtualized applications. The following parameters are used to determine sizing:

- The number of virtual machines running in the environment
- The vCPU to core ratio
- The target CPU Utilization
- Virtual machine memory allocation
- Virtual machine memory reservations.
- SimpliVity virtual controller resource overhead

The following are some key SimpliVity appliance compute configuration options:

- Single and Dual socket availability
- Configurable from 8 to 22 cores per socket
- Up to 1.5 TB of memory per appliance

When sizing compute, the virtual controller compute resource overhead must be taken into account. Table 6-2 outlines the SimpliVity Virtual Controller compute resource overhead for each size HPE SimpliVity appliance.

Table 6-2 Virtual Controller resources

SimpliVity appliance size	Virtual Controller memory	Virtual Controller CPU
Extra Small	64 GB	2
Small	70 GB	2
Medium	108 GB	4
Large	114 GB	4

Scale-out workloads

A SimpliVity cluster can start as small as a single SimpliVity appliance and grow based on VM requirements. For production workloads, a two-clustered SimpliVity appliance is recommended to ensure storage is highly available. A two-node SimpliVity cluster requires no 10 Gigabit Ethernet (GbE) networking. Nodes are directly connected for DVP traffic (Figure 6-3).

CHAPTER 6
HPE SimpliVity Hyperconverged Infrastructure

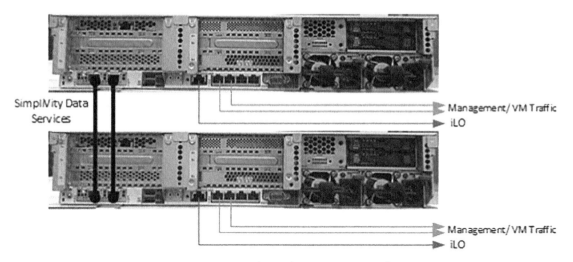

Figure 6-3 HPE SimpliVity direct connect appliances

Three or more nodes in a cluster require 10 GbE switching for data services traffic. All network traffic can be converged on the 10 GbE adapters. If networking is converged the management, vMotion, virtual guest networks, other networking requirements, and SimpliVity data services should be logically separated.

Adding a SimpliVity appliance to an existing cluster will scale-out both storage and compute. The addition of a SimpliVity appliance to an existing cluster does not impact the availability of the environment. The new appliance is deployed, and the available compute and storage resources are expanded within the cluster.

Compute-only scaling

With the efficiency provided by the inline deduplication and compression of the DVP, the need to scale compute resources often outpaces the need to scale storage capacity or performance. The DVP can be consumed by external hypervisor nodes to provide additional CPU and memory to support the environment. SimpliVity data stores are presented to external hypervisor hosts using 10 GbE Network File System (NFS.) This also allows for the integration of existing servers, blade or rack mount, into an HPE SimpliVity environment.

Once a SimpliVity datastore is presented to a hypervisor host providing compute resources the workloads stored on the datastore backed by the DVP will have access to all SimpliVity features, functions, and benefits. Even though the compute state of the workloads is not running on a SimpliVity host, they are accelerated and protected just as if the workloads were running on a SimpliVity appliance.

Resiliency

The architecture of the HPE SimpliVity infrastructure minimizes single points of failure by providing component-level protection in a node, synchronous data protection in a SimpliVity cluster, policy-based local backups, and asynchronous policy-based data protection between SimpliVity clusters.

Data is protected within a node using Redundant Array of Inexpensive Disk (RAID) with the HPE SmartArray controller. RAID provides disk-level data protection using RAID5 or RAID6 dependent on the size of the HPE SimpliVity appliance.

Data is also protected across nodes using Redundant Array of Inexpensive Node (RAIN). Using RAIN the data protected per virtual machine by synchronous replicating data across SimpliVity appliances within a cluster.

Policy-based local backups provide point-in-time protection and the ability to quickly recover full workloads or individual files/folders to a SimpliVity Cluster. These point-in-time backups are immutable and synchronously protected across nodes in SimpliVity cluster.

Finally, data can be protected Cross Cluster using SimpliVity's efficient cross-site replication. This provides asynchronous policy-based replication between clusters across a SimpliVity federation.

Hardware inventory

This is an example bill of materials for 4 HPE SimpliVity medium appliances. Each appliance is configured with dual Intel Xeon Silver 4114 processors and 576 GB of memory.

Table 6-3 4 HPE SimpliVity mediums: Bill of materials

Qty	Product #	Product description
4	Q8D81A	HPE SimpliVity 380 Gen10 Node
4	Q8D81A 001	HPE SimpliVity 380 Gen10 VMware Solution
4	826850-L21	HPE DL380 Gen10 Intel Xeon-Silver 4114 (2.2 GHz/10-core/85 W) FIO Processor Kit
4	826850-B21	HPE DL380 Gen10 Intel Xeon-Silver 4114 (2.2 GHz/10-core/85 W) Processor Kit
4	826850-B21 0D1	Factory integrated
8	Q8D86A	HPE SimpliVity 288G 12 DIMM FIO Kit
4	Q5V87A	HPE SimpliVity 380 for 6000 Series Medium Storage Kit
4	873209-B21	HPE DL38X Gen10 x8/x16/x8 PCIe NEBS Riser Kit
4	873209-B21 0D1	Factory integrated
4	P01366-B21	HPE 96W Smart Storage Battery (up to 20 Devices) with 145 mm Cable Kit

(Continued)

CHAPTER 6
HPE SimpliVity Hyperconverged Infrastructure

Table 6-3 4 HPE SimpliVity mediums: Bill of materials—cont'd

Qty	Product #	Product description
4	P01366-B21 0D1	Factory integrated
4	804338-B21	HPE Smart Array P816i-a SR Gen10 (16 Internal Lanes/4 GB Cache/SmartCache) 12G SAS Modular Controller
4	804338-B21 0D1	Factory integrated
4	652503-B21	HPE Ethernet 10 Gb 2-port 530SFP Adapter
4	652503-B21 0D1	Factory integrated
4	700751-B21	HPE FlexFabric 10 Gb 2-port 534FLR-SFP+ Adapter
4	700751-B21 0D1	Factory integrated
8	830272-B21	HPE 1600W Flex Slot Platinum Hot Plug Low Halogen Power Supply Kit
8	830272-B21 0D1	Factory integrated
4	BD505A	HPE iLO Advanced 1-server License with 3yr Support on iLO Licensed Features
4	BD505A 0D1	Factory integrated
4	Q8A62A	HPE OmniStack 8-14c 2P Medium SW
4	733664-B21	HP 2U Cable Management Arm for Easy Install Rail Kit
4	733664-B21 0D1	Factory integrated
4	867809-B21	HPE Gen10 2U Bezel Kit
4	867809-B21 0D1	Factory integrated
4	826703-B21	HPE DL380 Gen10 SFF Systems Insight Display Kit
4	826703-B21 0D1	Factory integrated
4	733660-B21	HP 2U Small Form Factor Easy Install Rail Kit
4	733660-B21 0D1	Factory integrated
4	826706-B21	HPE DL380 Gen10 High Performance Heat Sink Kit
4	826706-B21 0D1	Factory integrated
1	H1K92A3	HPE 3Y Proactive Care 24x7 SVC
4	H1K92A3 R2M	HPE iLO Advanced Nonblade—3yr support
4	H1K92A3 Z9X	HPE SVT 380 Gen10 Node (1 Node) Support
4	H1K92A3 ZA6	HPE OmniStack 8-14c 2P Medium Support
1	HA114A1	HPE Installation and Startup SVC
4	HA114A1 5LY	HPE SimpliVity 380 HW Startup SVC
1	HA124A1	HPE Technical Installation Startup SVC
4	HA124A1 5LZ	HPE SVT 380 for VMware Remote SW St SVC
3	HF385A1	HPE Trng Credits ProLiant/HybridIT Svc

This example will provide the following resources if all nodes are deployed in a single cluster:

Total physical CPU cores: 80 @ 2.2 GHz

*Usable CPU Resources: 140 GHz

*Usable Memory Resources: 1872 GB

Total Effective Storage Capacity: ~49 TB

(* Usable after accounting for OVC resource overhead)

Summary

The HPE SimpliVity hyperconverged infrastructure is powered by the HPE SimpliVity Data Virtualization Platform to provide enterprise efficiency, built-in resiliency, native data protection, and flexible scale. These integrated features simplify management of resources within the infrastructure and reduce risks associated with data protection and infrastructure resiliency.

HPE SimpliVity's appliance architecture provides design flexibility to support a variety of enterprise use cases including Tier-1 Applications, Datacenter Consolidation, Private Cloud/Hybrid Cloud, and Virtual Desktop Infrastructure (VDI). Across all use cases, HPE SimpliVity provides efficiency and reduces complexity, which in turn improves productivity and reduces costs.

7 HPE Synergy and Composable Infrastructure

INTRODUCTION

This chapter covers an upgrade from aging and static hardware to new and advanced infrastructure used to support a strategically important application then needs to be highly available. The solution includes multiple workloads and applications based on HPE Synergy referred to simply as Synergy throughout this chapter.

Before the upgrade, a large portion of the current application platform was implemented on generation 6 and 7 HPE rack mount servers. These servers were attached to an old storage platform that needed to be replaced along with the compute portion of the environment. Virtualization is the foundation for all applications running in this environment.

Along with the aging infrastructure concerns, improving resiliency was an important aspect of the upgrade. The application platform that was refreshed was considered both business and mission-critical for the infrastructure owner and the platform users. If an outage occurred, then the ability for outside users to connect to the application platform would cause serious business disruptions.

Why Synergy?

After evaluating the various options, a Composable Infrastructure was determined to be the ideal solution for bridging the gap from the current bare-metal infrastructure to an infrastructure that could support all workloads. Synergy supports all the potential use cases and critical success factors. Some key characteristics, making it the ideal platform for such an upgrade, are described below:

- **Resiliency**

    The application platform that was refreshed is both mission- and business-critical. The goal of the refreshed infrastructure is to reduce the "blast radius" of a failure. Blast radius is the amount of infrastructure that can be affected by a single failure. One example of limiting blast radius is putting infrastructure on separate racks with separate power. Later in the chapter, it will be explained how Synergy was designed to be a highly available and redundant platform. Combined with virtualization technologies, such as moving VMs and creating back up procedures, the new infrastructure was planned to be more resilient then in the past.

- **Management ease of use**

  In the existing environment, there was not a comprehensive management tool or strategy for the infrastructure. Synergy uses HPE OneView (covered in Chapter 11) as the management tool to control all aspects of the Infrastructure settings. The software-defined capabilities of OneView ease the numerous management pain points for the user including firmware updates and configuration changes.

- **East of deployment**

  The business generated by the current application platform is a growing portion of this environment. It is scaling rapidly that calls for a lot of new infrastructure quickly. With the legacy infrastructure, the process of deploying new servers was very time-consuming. The capability of OneView to use templates streamlines this process. Auto-integrating hardware into the management environment also allows the infrastructure to be deployed as soon as it is received.

- **Performance**

  The highly integrated environment of Synergy resulted in higher performance for this application platform.

Synergy overview

This section covers some details of Synergy and then we will come back to the specific solution for the application platform upgrade.

Composable infrastructure is the next logical step in the evolving IT landscape as shown in Figure 7-1. The left of the figure depicts infrastructure that was built in silos. Servers were separate from storage and each had their own networking. The next step was to merge these technologies together to make them easier to use and build. Converged infrastructure was prebuilt and integrated infrastructure was validated to work together. It bundled the silos of infrastructure. Hyperconverged came next in the evolution, combining compute and storage into a single appliance. These appliances consist of industry standard servers with a large amount of disk space that could be shared across each other via software-defined storage. It is a great fit for specific workloads though such as virtualization and VDI. Composable Infrastructure is the latest evolution and removes the "siloed" approach to IT. Figure 7-1 shows this evolution in full.

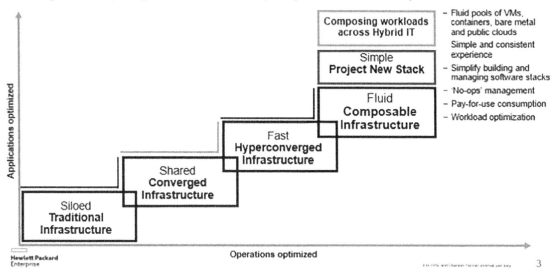

Figure 7-1 Hybrid IT progression

HPE composable infrastructure

The goals of composable infrastructure are to run anything, move faster, work efficiently, and unlock value. In order to achieve these goals, composable infrastructure has three architectural design principles. Those principles are fluid resource pools, software-defined intelligence, and a unified API as depicted in Figure 7-2.

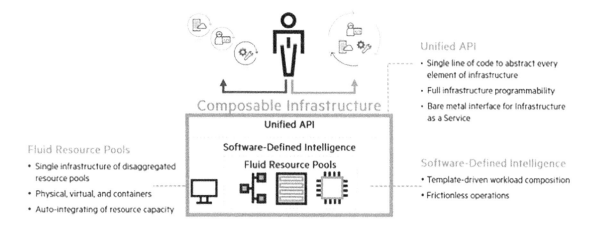

Figure 7-2 Composable infrastructure

Traditionally, infrastructure was silos of servers, storage, and networking. When a new project emerged, deployment typically took a long time because all of the components in these silos needed to be built and connected. This typically led to overprovisioning because project managers and administrators did not want to complete the whole cycle again for a new capacity. With Composable Infrastructure, fluid resource pools combine servers, storage, and networking. They can then start as a stateless resource and be combined/configured to fit any need quickly. As more infrastructure is added into the environment it is autointegrated and added to the pool. Resources then can be decommissioned as well be used for separate use-cases and based on need. By sizing the infrastructure properly and having the ability to expand/contract on demand, overprovisioning can be reduced, lowering both CAPEX and OPEX.

In traditional silos, each component also had its own management interface for administrators to provision and support their own infrastructure. Changes to the environment meant coordination across multiple teams and tools. This stretched project timelines and delayed deployments. Composable infrastructure implements an environment that is delivered across a single interface that can compose and recompose the infrastructure as needed. Templates are used so that provisioning is repeatable and changes are rapid. Without manual intervention, error is reduced, resulting in less downtime.

Since every management tool was a silo in the past, each had its own API or scripting tools for the components it managed. This means that integrating and automating pieces of infrastructure was very complex and often required a multitude of experts. With Composable Infrastructure all infrastructure is controlled via a single unified or RESTFUL API that integrates with the single management interface. Through this, a single line of code can reference a template that can be leveraged to provision infrastructure required for an application. Old workflows can be eliminated by automation, leading to faster deployments and quicker results. Third-party management tools can also take advantage of the hardware as software by building integration into their own tools.

The next sections will cover some of the key components of the Synergy hardware.

Synergy 1200 frame and components

The Synergy 12000 frame is a 10U chassis that was built on industry standards. It fits easily into existing and new datacenters. The front of the frame gives you the flexibility to add a multitude of device modules. There are half-height, full-height, and double-wide sized modules for compute and storage. The frame can hold a maximum of 12 half-height single width device modules. There are also two appliance bays apart from the compute and storage slots on the side of the frame. These bays hold management modules such as the composer that hosts OneView. The back of the frame hosts the networking components, power, and cooling. There are six redundant power supplies, ten redundant fans, two redundant frame link modules (for air-gapped management), and three redundant network fabrics. Figure 7-3 shows the front of a Synergy frame.

Figure 7-3 Synergy frame

Scaling within the frame is achieved by adding additional device modules. Scaling within the rack is as easy with auto-integration into OneView as you make the frame's management connections. Linking the frame to the existing frames' management network is all that needs to be done. As the frames scale across the data center, all the frames get connected to the same management network so that they are all in a single management interface.

The Synergy frame was built with extensive redundant architecture to eliminate single points of failure. Every component is built with that idea in mind. Even the mid-plane, which can provide 16Tb of bandwidth to device modules, is completely passive. It is also "photonics ready" meaning that, in the future, it can use light and has the capability of being orders-of-magnitude the speed of copper. The Synergy 12000 frame was designed to easily integrate into today's data center, to efficiently scale, to be highly redundant, and to be future-proof for years to come.

Composer

The HPE Composer is the management appliance that fits into one of the two side appliance bays, which are unique, do not take up any fluid resource pool space, and you only need one Composer for 21 frames. The Composer hosts HPE OneView that is the management and automation engine for all infrastructure within Synergy. It allows the users the capability to template hardware, set up networking, update firmware, monitor alerts, and much more. It is built on a RESTFUL API so that anything that can be done in the web GUI can be done programmatically. It also allows for third-party software manufacturers to provide integrations. Some examples of suppliers that already have made integration today are VMware, Microsoft, Chef, Puppet, and Ansible. Figure 7-4 shows the Composer in a Synergy frame.

CHAPTER 7
HPE Synergy and Composable Infrastructure

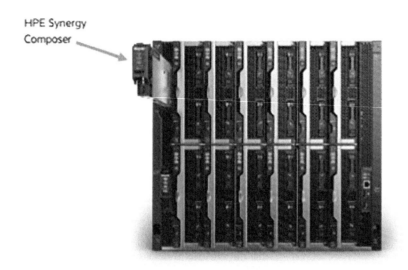

Figure 7-4 Composer in a Synergy frame

A single pair of redundant composers support up to 21 frames or a maximum of 252 compute modules in a data center. They are all connected through the management ring built by the frame link modules in the back of the frame. This is the same management network that provides the auto-integration capability.

Image Streamer

The HPE Image Streamer is another management appliance that can go into one of the two side appliance bays of a Synergy frame. It provides operating system provisioning for the compute modules within multiple frames. Image Streamer can also capture, edit, and store your golden images for use. The goal of Image Streamer is to eliminate inefficient procedures related to getting an operating system up-and-running. Image Streamer is shown in Figure 7-5.

Figure 7-5 Image Streamer in a Synergy frame

A pair of redundant Image Streamers is required for each fabric domain. The Synergy fabric architecture eliminates the need for a top of rack switch, reducing latency between servers and improving performance in the data center by leveraging pass-thru satellite modules and redundant master modules. Not only is this fabric architecture reliable, but it also simplifies setup, delivers better performance, and performs at high speed. It works through both the management network to integrate into OneView along with the data network to actually provision the operating systems. The operating systems are provisioned by an iSCSI boot, which is fully automated and requires no management setup. When you select an operating system deployment in OneView the golden volume selected is copied. Plan scripts personalize the OS volume, and the composer automatically configures the iSCSI boot. From there, the compute module is provisioned and ready to be used much quicker than it would via a traditional method.

Compute modules

Synergy compute modules are built for every workload from general purpose to mission-critical. They support the full range of Intel processors, the full memory compliment, along with flexible networking and storage capabilities. There are four models available at the time of this writing including the SY480 Gen 10, the SY660 Gen 10, the SY620 Gen 9, and the SY680 Gen 9. The SY480 Gen 10 is a half-height single width two processor compute module. It supports the majority of workloads and is the most deployed server in Synergy. It can support up to 56 processor cores, 3 TB of memory, and has two storage slots. The SY660 Gen 10 is a full-height single width four processor compute module. It is ideal for scale-up data-intensive workloads with double the memory and storage compared to the SY480 Gen 10. The SY480 Gen9 is Xeon E5, the SY620 and SY660 Gen9 are E7, and the Gen10 compute modules are Skylake processors.

Figure 7-6 depicts the compute modules supported by Synergy.

CHAPTER 7
HPE Synergy and Composable Infrastructure

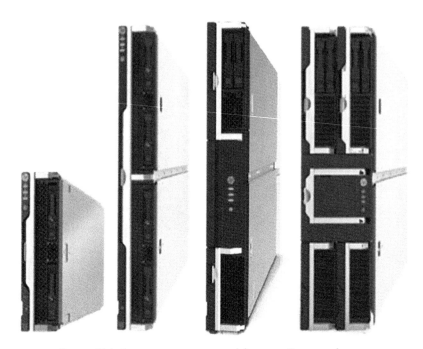

Figure 7-6 Synergy compute modules in a Synergy frame

While the SY480 Gen 10 and the SY660 Gen 10 are more for general-purpose workloads, the SY620 Gen 9 and the SY680 Gen 9 are for mission-critical workloads. They are built with the Intel EX processor family and can support up to four processors and 6TB of memory in a full-height single width or double width form factory. They are built to support workloads that require high availability such as advanced analytics or business processing. All these compute modules can be mixed into a single Synergy frame to support any workload.

Storage

When it comes to storage within Synergy, the key is flexibility. There are a number of solutions including direct-attached storage (DAS), software-defined storage, and traditional fiber channel attached storage. Depending on the workloads, a single Synergy frame can support multiple storage solutions. Figure 7-7 shows a Synergy storage module and a 3PAR frame.

Figure 7-7 Storage in a Synergy frame

DAS has two different options in Synergy. The first is on the compute module itself. For each compute module, there is a choice for local drives to use for boot or additional data. Those drives can be HDD, SSD, NVMe, or uFF drives. If no local drives are needed, there is even a stateless model that removes the drive slots on the compute module. The second way to connect direct-attached storage to a compute module is through a D3940 storage module. The D3940 is a half-height double width module that can support up to 40 SFF drives. The drives can be direct-attached to any compute module within the frame with the D3940 is installed. The connections are made through a non-blocking SAS fabric that can support upwards of 2M IOPS. When using Gen10 Compute Modules a maximum of five of these storage modules can be put into a frame for a total of 200 SFF drives in a single frame and with Gen9 a maximum of four of these modules are supported.

By utilizing software direct-attached storage can be treated as a shared storage platform and be leveraged by compute modules across frames. Synergy supports a number of solutions including StoreVirtual VSA and VMware vSAN. This eliminates the additional cost and complexity of traditional storage and storage networking. If software-defined storage is not the solution, Synergy also supports traditional fiber channel SANs as descried in the high-speed SAN chapter. The Virtual Connect CNAs, built to support Ethernet and FC, or a purpose-built HBA can be leveraged to make the connection to a SAN. In this way, existing storage platforms such as HPE 3PAR can be leveraged.

Networking

There are two distinct networks in Synergy. The first is the management network that was discussed earlier in the chapter. At the back of the frame on the side opposite the appliance bays there are slots filled by what are called frame link modules. Frame link modules allow the network connections to the management modules of Synergy. They are completely air gapped from the data plane to provide resiliency during a network outage. Their usage is twofold: one is to create a management ring between frames for auto-integration and scaling and the second is to provide uplinks to the top of the rack management switches.

Figure 7-8 depicts Synergy networking.

CHAPTER 7
HPE Synergy and Composable Infrastructure

HPE Synergy Master and Satellite modules

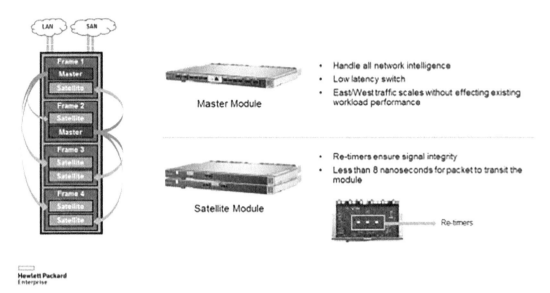

Figure 7-8 Networking in a Synergy frame

The other network is the data network that carries application traffic. Synergy provides three separate and redundant network fabrics for this purpose. Each frame contains three pairs of interconnect module slots for a variety of redundant network options including Virtual Connect switches and fiber channel switches.

For the data, a master/satellite architecture is implemented. In the past with older bladed infrastructure, each chassis needed its own network switches. With the master/satellite architecture, now a pair of redundant master switches can support up to five Synergy frames. The master modules provide the uplinks to the top of rack or end of row switches, while the satellite modules provide connections back to the master modules. There is no recognized latency or oversubscription from the satellite modules to the master modules. Both master modules can take use of M-LAG and be treated like a single switch. This architecture reduces complexity and makes scaling much easier and cheaper.

Solution architecture

This section describes the solution architecture of the refresh for the application platform discussed earlier in the chapter.

Hardware components

The following is a description of the hardware components for this solution:

- Two Synergy 12000 Frames in each frame with:
  - Six SY480 Gen 9 Compute modules each with:
    - Two E5-2699v4 processors
    - 512GB DRAM
    - Two 600GB HDD controlled by P240nr
    - 3820C 10/20Gb CAN for network connectivity
  - One VC SE 40Gb F8 Master Module each with:
    - Four QSFP+ to SFP+ transceivers populated with:
      - Two 10Gb SFP+ transceivers
      - Two 8Gb FC SFP+ transceivers
    - One 40Gb QSFP+ to QSFP+ DAC Cable
  - One Synergy 20Gb Interconnect populated with:
    - Two 12x10Gb AOC cables
  - One HPE Composer
  - Two Frame Link Modules each with:
    - Two CAT6 cables (one for management ring, one for management uplink)
  - Six 2650W AC or DC power supplies (depending on site)
- Two 3PAR 8440 2 Node systems with:
  - 24 400GB SSD
  - 2 4 port 16Gb FC uplinks
  - All-inclusive Software
  - AC or DC power supplies (depending on site)

CHAPTER 7
HPE Synergy and Composable Infrastructure

- Two Brocade Switches
    - 6510 24P 8Gb for DC sites
    - SN3000 24P 8Gb for AC sites

Figure 7-9 shows a front view of this equipment in racks.

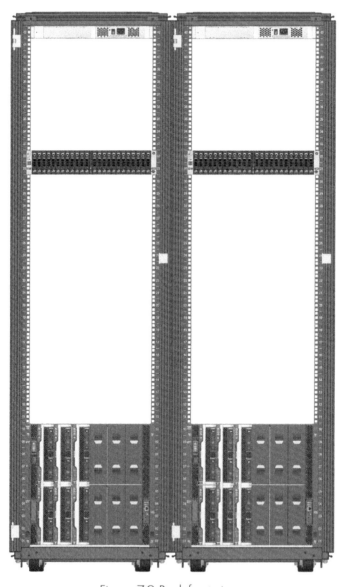

Figure 7-9 Rack front view

Figure 7-10 shows that back view of the rack.

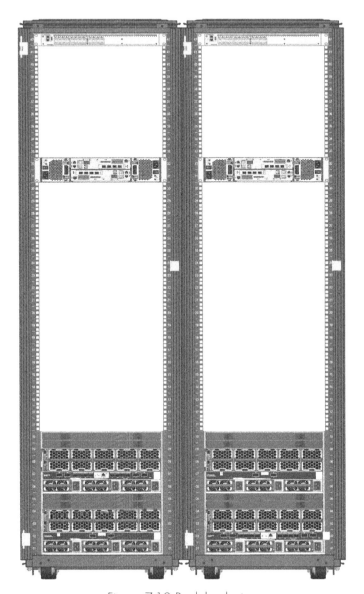

Figure 7-10 Rack back view

CHAPTER 7
HPE Synergy and Composable Infrastructure

Summary

This chapter demonstrates how HPE Synergy is the ideal platform for a composable infrastructure solution. The upgrade environment described in this chapter supports all workloads and takes the burden off IT administrators by providing a flexible and fluid pool of IT resources that can be provisioned as software. Synergy supports both orchestrating new VMs as they are required as well as reclamation of resources once an application is removed. It is the ideal platform to foster innovation within the data center, enabling IT administrators to spend less time managing their environment and more time driving new products and services.

8 HPE HANA TDI Solution and Superdome Flex

INTRODUCTION

There are many options for users who deploy SAP's HANA Database environment. These options include on-premises, private cloud, or public cloud deployment. For those users who choose an on-premises deployment model there is yet another choice to make, HANA appliance or a HANA Tailored Datacenter Integration (TDI) approach.

Until a few years ago SAP HANA was only delivered as a fully integrated appliance that included HPE's CS500 and CS900 appliances. As HANA and client expertise matured, SAP offered the flexibility of a TDI deployment. TDI allows users the flexibility to standardize on their existing reference architecture for hardware, network, and storage components for HANA deployments.

The SAP HANA workload described in this chapter shows a successful TDI deployment that derived the common benefits of this approach including:

- Lower TCO by leveraging existing storage and network components
- Standard technology stack
- Simplified management with existing tools

Solution requirements

This HANA scale up workload was a result of the migration from SAP Enterprise Central Component (ECC) to SAP's Suite on HANA (SoH.) Although SAP HANA Business Suite has the HANA database, the same SAP ECC modules are present. The resulting benefit is the significant performance boost of SAP HANA at the database layer without restructuring the application layer.

Business requirements

- Provide a solution that supports rapid business growth with a scalable HANA platform.
- Increase productivity of SAP end-users with faster response times within SoH solution.
- Provide mission-critical platform to support the workload 24x7 worldwide operations.

Technical requirements

- Provide a scalable HANA solution capable of scaling to beyond four sockets and handling certified SAP HANA workloads to 8-12 TB.

- Provide a High Availability (HA) clustering solution capable of providing local failover without any transaction loss in SAP HANA environment.

- Provide Disaster Recovery (DRI) solution for new SoH environment.

Solution overview

HPE's Superdome X was designed and engineered as a highly scalable and reliable server platform for scale-up HANA workloads. At the time this HANA workload was deployed, the Superdome X was capable of scaling to 16-sockets and 16 TB configurations for HANA as shown in Figure 8-1.

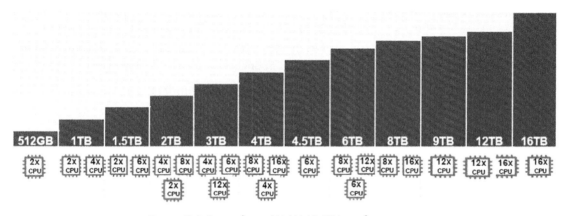

Figure 8-1 Superdome X HANA TDI configurations

The Superdome X allows this HANA workload to scale within the deployed Superdome X enclosure. The result is a less disruptive growth path for the HANA workload and minimal risk to the overall availability of the HANA workloads in the user's environment.

The HPE Superdome X was deployed in a multi-partition configuration. The Superdome X has the ability to create electrically isolated hard partitions, called nPartitions, within the Superdome X enclosure. The result is fully isolated compute nodes for HANA workloads.

The SAP HANA SoH architecture for this workload consisted of Qty (2) Superdome X enclosures. These enclosures consisted of both production and nonproduction Superdome units. Both Superdome X unites were partitioned into multiple SoH instances. The Production SoH instance consisted of Qty (3) Superdome 2-socket Cellbades and 4.5 TB of Memory. This SoH instance was also configured into an HPE Serviceguard for Linux cluster with the Quality Assurance (QA) SoH instance. Serviceguard software facilitates failover between hard partitions (nPartition or nPars for short) and systems. This is referred to as a Dual Purpose HANA HA configuration. In the event of a failure of the Production nPartition or enclosure, the SoH QA instance would be shutdown on the nonproduction Superdome X and the production SoH instance would be started. The result is a complete automated failover of the production SoH instance bewteen physical Superdome X enclosures without a dedicated HA failover partition.

The development SoH instance is located in the nonproduction Superdome. It consisted of Qty (2) 2-socket Cellblades and 3 TB of memory. There was an additional requirement to the HANA Serviceguard cluster. The previous ECC environment consisted of both production and development Serviceguard clusters to allow for any changes such as version updates, operating system updates, patches, and so on. These will first be tested and validated in the development cluster before applying them to production. The user requested this requirement be designed into the new SoH architecture. As a result, there is a second partition (nPar) in the production Superdome X for development Serviceguard cluster failover.

The Superdome X HANA architeture is detailed in Figure 8-2.

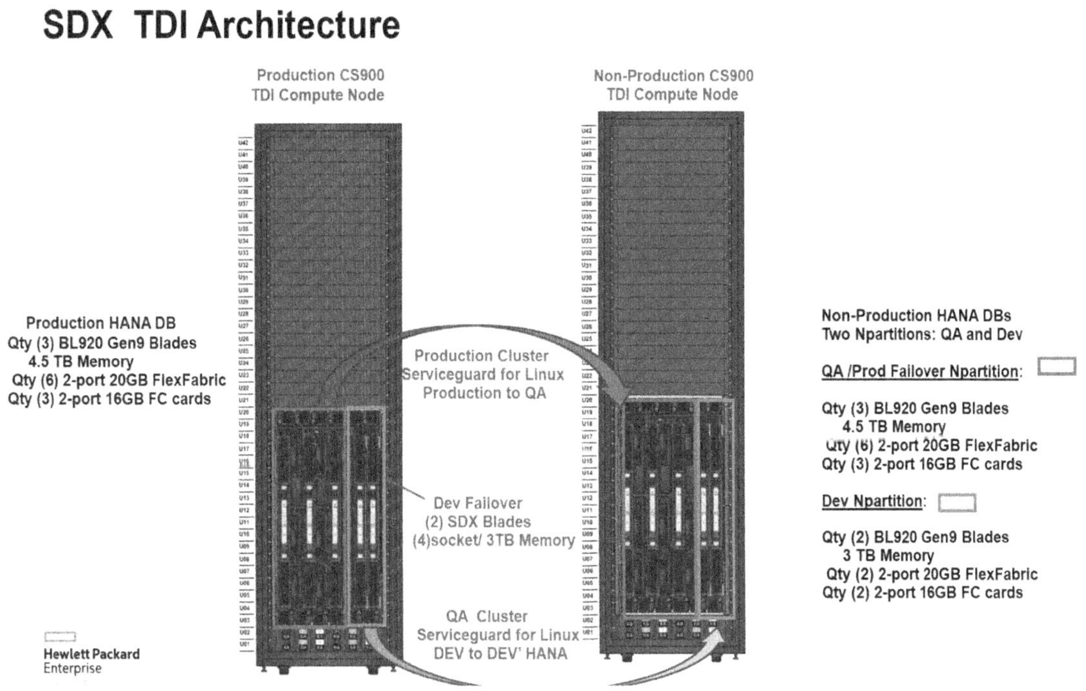

SDX TDI Architecture

Figure 8-2 Superdome X SAP HAA SoH Architecture

CHAPTER 8
HPE HANA TDI Solution and Superdome Flex

In order to achieve the objectives listed above, the Superdome X HANA TDI-based solution shown in Figure 8-2 was deployed with the following key components per server:

- Qty (2) HPE Superdome X Server with the following characteristics:
- Qty (5) HPE Superdome X Server 2-socket Cellblades with intel E7-8880v4 (22) cores X processors
- 1.5 TB RAM per Cellblade/ 7.5TB RAM total per SDX
- Qty (4) Ethernet 10GB Dual-port 560FLB Network adapter
- Qty (4) 16GB Fiber Channel SAN adapters
- Qty (2) Brocade 16GB SAN switches
- Qty (2) HPE 6125XLG Network switches
- HPE Proactive Care Support for 3 years
- HPE HANA Center of Expertise (COE) access

How did we arrive at this solution?

SAP HANA database sizing is based on the data stored in main memory. The HANA data is also compressed, so it is critical the appropriate SAP sizing methodologies are followed to determine the size of the HANA TDI server. In the case of the existing ECC environment, SAP note 1793345 provides the necessary steps and the SAP HANA ABAP (SAP programing language) sizing script that is run against the existing ECC database that generates a report for the SoH sizing. The ABAP report output for this particular SoH workload TDI is shown in Figure 8-3.

```
------------------------------------------------------------------------
| SIZING RESULTS IN GB                                                 |
|----------------------------------------------------------------------|
| Based on the selected table(s), the anticipated maximum requirement are |
|                                                                      |
| for Suite on HANA:                                                   |
| - Memory requirement                                         3.915,3 |
| - Net data size on disk                                      2.203,7 |
|                                                                      |
| - Estimated memory requirement after data clean-up           3.462,9 |
| - Estimated net data size on disk after data clean-up        2.203,7 |
|                                                                      |
| Other possible additional memory requirement:                        |
| - for an upgrade shadow instance                               231,5 |
------------------------------------------------------------------------
Check the FAQ document attached to SAP Note 1872170 for explanations on how
to interpret the sizing terms and calculations.

Sizing report:                                            ZNEWHDB_SIZE
Version of the report:                                             61a
Date of analysis:                                           05.08.2016
Selected accuracy:                                                   M
Number of work processes used:                                      08

SID                                                                ECP
NW release:                                                   701 SP 8
Kernel version                                             721_EXT_REL
Operating system on AS                                 HP-UX B.11.31 U i
Type of analyzed database:                                      ORACLE
Database version:                                           11.2.0.2.0
Unicode system:                                                    Yes
Used size on disk of the analysed database in GB:             7731.1

Number of tables successfully analyzed:                       82.366
Number of tables partially analyzed:                               0
Number of tables with error:                                       0

------------------------------------------------------------------------
| MEMORY SIZING CALCULATION DETAILS                     HANA SIZE IN GB |
|----------------------------------------------------------------------|
|   Column store data                                          1.867,6 |
| + Row store data                                                41,7 |
|   -------------------------------------------------                  |
| = Anticipated memory requirement for the initial data        1.909,3 |
| + Cached Hybrid LOB (20%)                                       46,7 |
| + Work space                                                 1.909,3 |
| + Fixed size for code, stack and other services                 50,0 |
|   -------------------------------------------------                  |
| = Anticipated initial memory requirement for HANA            3.915,3 |
------------------------------------------------------------------------
```

Figure 8-3 Sample SAP ABAP sizing report

This ABAP report was run on the existing ECC which consisted of HP-UX 4-socket blades model BL870i4. The ECC landscape is depicted in Figure 8-4.

SAP ECC Database Environment

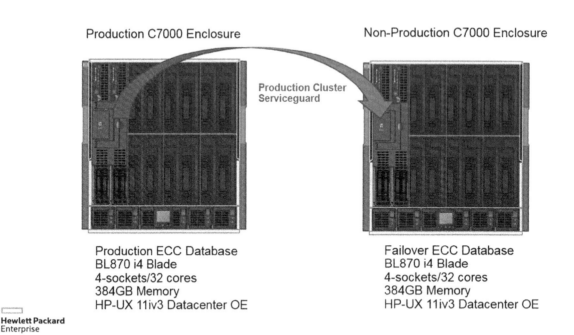

Figure 8-4 Existing SP ECC production environment to be replaced

The sizing resulted in the stated HANA Database size of 3.9 TB for SoH. The Superdome X has defined nPartition and memory sizes that are supported. Figure 8-5 details the available memory and partitions sizes available with 32 GB memory Dual Inline Memory Modules (DIMMs.)

The 4.5 TB size was chosen based on the sizing requirement 3.9 TB. With anticipated first year growth, the sizing was rounded to 4.5 TB for the production SoH Database. Figure 8-5 shows the Superdome X-supported memory configurations.

nPartition memory configurations using 32GB DIMMs					
Sockets	Blades	Memory	DIMM size	# DIMMs	Note
16	8	12TB / 12288GB	32GB	384	
16	8	8TB / 8192GB	32GB	256	
16	8	4TB / 4096GB	32GB	128	
12	6	9TB / 9216GB	32GB	288	
12	6	6TB / 6144GB	32GB	192	
12	6	3TB / 3072GB	32GB	96	
8	4	6TB / 6144GB	32GB	192	
8	4	4TB / 4096GB	32GB	128	
8	4	2TB / 2048GB	32GB	64	
6	3	4.5TB / 4608GB	32GB	144	
6	3	3.0TB / 3072GB	32GB	96	
6	3	1.5TB / 1536GB	32GB	48	
4	2	3TB / 3072GB	32GB	96	
4	2	2TB / 2048GB	32GB	64	
4	2	1TB / 1024GB	32GB	32	
2	1	1.5TB / 1536GB	32GB	48	
2	1	1.0TB / 1024GB	32GB	32	
2	1	0.5TB / 512GB	32GB	16	

Figure 8-5 Memory configurations using 32GB DIMMs

In addition to the (5) Cellblades/7.5 TB of memory per Superdome X, there were several new infrastructure changes made with the HPE HANA TDscale-up workload deployment.

These include the following:

- Network interfaces on the Superdome X are 10 Gigabit(Gb) Network Interface Cards (NICs) for support of required high-speed connections for HANA replication, end-user connectivity, and HANA backup

- The Fiber Channel (FC) Storage Area Network (SAN) infrastructure was upgraded to 16 GB FC including 4 X Qlogic 16 GB dual-port FC Host Bus Adapters (HBAs) and a pair of 16 GB Brocade FC switches in the Superdome X Interconnect bays.

- A new HPE 3PAR array for support of the SoH deployment. This was an HPE 3PAR 8400 with 4 nodes, Qty (56) x 7.68 TeraByte (TB) Small Form Factor (SFF) Solid-State Disk (SSD) drives and Virtual Copy software.

Hardware inventory

Table 8-1 shows the Bill of Materials (BOM) for this HANA TDI workload solution.

Table 8-1 BOM for SDX TDI solution

SAP HANA TDI - Superdome X Solution		
		Hardware
1	M0S67A	HPE 1075mm Shock Intelligent Rack
1	P9H73A	HPE SDX for SAP HANA Base Encl
2	787635-B21	HP 6127XLG Blade Switch Opt Kit
2	C8S47A	Brocade 16Gb/28c PP+ Embedded SAN Switch
3	P9H60A	HPE BL920s Gen9 E7-8880v4 44c Svr Blade
36	P9H75A	HPE SDX DDR4 128GB (4x32GB) Mem Module
6	P9H77A	HPE FF 20Gb 2p 650FLB Adptr for CS900
3	P9H78A	HPE QMH2672 16Gb FC HBA for CS900
1	P9H56A	HPE SDX SAP HANA TDI Scale-up Soln
8	C7536A	HP Ethernet 14ft CAT5e RJ45 M/M Cable
1	AF528A	HP 5xc 13 PDU Extension Bars Kit
8	AF520A	HPE Intelligent 4.9kVA/L6-30P/NA/J PDU
1	M0S66A	HPE Virtual Rack
1	P9H63A	HPE CS900 SAP HANA SG Primary Config
		Software
3	J7J28A	HPE CS SAP HANA iLO Adv-BL 3yr TSU LTU
1	J7J23A	HPE CS SAP HANA Advanced Partitions LTU
3	N0U73A	SLES SAP 2Skt/1-2 VM 3yr 24x7 Flx LTU
6	P9B46A	HPE SG Ent for SAP HANA 1Y24x7 Flx SW
		Support
1	H1K92A3	HPE 3Y Proactive Care 24x7 Service
2	H1K92A3#9LS	HPE B-Series 8/24c Switch PowerPK Supp
2	H1K92A3#SPM	HPE 612x Blade Switch Support
3	H1K92A3#TPZ	HPE CS iLO Adv 3yr TSU LTU Support
1	H1K92A3#XWT	HPE SDX for SAP HANA Base Encl Supp
3	H1K92A3#XWU	HPE BL920s Gen9 Svr Bld Supp
1	H1K92A3	HPE 3Y Proactive Care 24x7 Service
3	H1K92A3#YMJ	HPE SLES SAP 1-2 VM 3yr Flx LTU Support
1	H1K92A3	HPE 3Y Proactive Care 24x7 Service
6	H1K92A3#YNB	HPE SG Ent for SAP HANA Supp

(Continued)

Table 8-1 BOM for SDX TDI solution—cont'd

SAP HANA TDI - Superdome X Solution		
		Services
1	HA124A1	HP Technical Installation Startup SVC
1	HA124A1#5VH	HPE Startup Integrity Superdome X SVC
1	H8A03A1-122	HPE CS SAP HANA SG SW FE Depl SVC

Software inventory

There are several software components that are essential to the operation of this HANA TDI workload. Table 8-2 shows the software components that are implemented in this solution.

Table 8-2 Software to implement SoH workload

Superdome X HANA TDI Solution Software
– SLES (SUSE Linux Enterprise Server) for SAP Applications 12 SP3
– HANA Database (Licenses procured directly from SAP)
– HPE Serviceguard for Linux with SAP toolkit

Figure 8-6 shows that SLES for SAP Applications is a bundle of software and services designed to address the specific needs of SAP users.

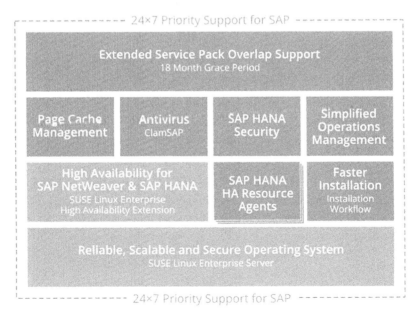

Figure 8-6 Software and services included in SLES for SAP Applications

The following is a list of highlights that are a part of SLES for SAP Applications.

- Linux kernel 4.4 is a part of SLES for SAP Applications that provides the latest performance optimizations for HPE server hardware. In addition, the kernel provides functions like **ftrace** that are used by SUSE Linux Enterprise Live Patching that enables a kernel patch to be applied without stopping SAP HANA and without requiring a reboot to resolve critical kernel vulnerabilities.

- The ability for administrators to perform patch and service pack rollout is provided through the use of the BTRFS filesystem along with integration in the package management libraries. Installing the root filesystem with BTRFS will enable the rollback capability.

- Like HPE Serviceguard, HA for SAP NetWeaver and SAP HANA can be used to provide failover and continuity for SAP deployments that can be leveraged in test and development scenarios supporting both physical and virtual deployments.

There are several installation methods included in SLES for SAP Applications that are available to Linux and SAP Basis administrators. The standard way is to use the Installation Workflow to automate the installation of the operating system and an SAP application. The workflow is divided into steps – the installation of the operating system, SAP Installation Wizard parts 1, 2, and 3. Most of these steps do not need to be run immediately after each other, which allows for flexibility. An example is installing the operating system along with SAP Installation Wizard part 1 that copies the SAP install media. Then creating disk images that can be copied to other systems before starting part 2, which collects the SAP application installation options and part 3, which starts the SAP Installer.

Simplified Operations Management in SLES for SAP Applications includes several features – SAP specific installation patterns, system tuning for SAP and storage encryption.

- Three installation patterns are included—SAP BusinessOne Server Base, SAP HANA Server Base, SAP NetWeaver Server Base—that simplify working with RPM package dependencies when preparing to install an SAP application.

- The system tuning application saptune will automatically and comprehensively tune a system as recommended by SAP for use with SAP S/4HANA, SAP NetWeaver, or SAP HANA/SAP BusinessOne. Through the use of tuned profiles, saptune can adjust operating system parameters to match recommendations from specific SAP Notes.

- Through the use of cryptctl, partitions can be encrypted on the client systems containing sensitive data that might be running in the cloud or hosted datacenters. For clients to decrypt a partition, a cryptctl server stores keys on a KMIP 1.3 compatible server. The client will make an RPC request to the KMIP server that records the request and delivers the partition key required for decryption.

SLES for SAP applications includes two important security features—the SAP HANA Firewall and the OS Security Hardening Guide for SAP HANA.

- SAP HANA Firewall extends SuSEFirewall2 by providing administrators additional options to accommodate SAP HANA properly.
- The SLES for SAP applications resource library which is at suse.com includes the latest version of the "OS Security Hardening Guide for SAP HANA."

ClamSAP integrates the ClamAV anti-malware toolkit into SAP NetWeaver and SAP Mobile Platform applications. ClamSAP is a shared library that links between ClamAV and the SAP NetWeaver Virus Scan Interface (NW-VSI). The version of ClamSAP shipped with SUSE Linux Enterprise Server for SAP Applications 12 SP3 supports NW-VSI version 2.0.

The Linux kernel swaps out rarely used memory pages to free memory for cache operations. Response time for a user can be poor when a SAP NetWeaver or SAP HANA page has been swapped out. Page Cache Management limits the amount of page cache the Linux kernel uses when there is competition between SAP application memory and page cache.

Extended Service Pack Overlap Support (ESPOS) allows customers to perform service pack migrations within 18 months. Migrations can be scheduled more easily and testing before a migration can be performed under lesser time constraints.

SUSE Linux Enterprise Server Priority Support for SAP Applications offers technical support for SUSE Linux Enterprise Server for SAP Applications directly from SAP based upon SAP Solution Manager. This offering provides seamless communication with both SAP and SUSE reducing complexity and lowering TCO through the "One Face to the Customer" approach.

More information and links to everything that has been highlighted can be found at https://www.suse.com/documentation/sles-for-sap-12/book_s4s/data/cha_s4s_about.html.

Implementation plan

The user deploying the Superdome X SoH TDI workload was using the corporate datacenter location for the SoH deployment. The SoH TDI deployment is a significant shift in implementation tasks as compared to a SoH appliance deployment that utilizes factory integration of OS, HANA DBs, and network settings. As a result these tasks become integral part of a TDI implementations plan.

Table 8-3 Simplified implementation plan

HANA planning workshop
– HANA TDI configuration planning
– Determine HANA configuration parameters and cabling requirements.
– HA planning
– Document implementation plan
Project management
– Project management Kickoff meeting to define roles and resources.
– Provide status reporting
– Provide issue management
Hardware deployment
– Confirm Datacenter and Rack preparation complete
– Superdome X power up and diagnostic testing complete
– Confirm connectivity to Superdome X iLO console complete
SAP HANA installation
– Configuration of Superdome X embedded 6125 network switches
– Validate Network connectivity
– Configure embedded Brocade 16GB SAN switches
– Startup and configuration of 3Par storage array
– Configure management server
– Startup and initial configuration of Qty (2) Superdome X servers
– Configure Superdome X Npartitions to specifications.
– Install and Patch RHEL OS instances on all Npartitions
– Install and Patch HANA DB on SLES for SAP OS instances
– Create and configure HANA Instances
– Install SAP HANA studio on target PC or workstation
– Import sample test data into HAAN Databases
– Run KPI scripts on production server
– Document Superdome X and HANA configurations
HPE Serviceguard for Linux installation
– Configure SAP replication for use with Serviceguard for Linux
– Configure Serviceguard Packages for SAP HANA DB
– Complete Serviceguard / HANA testing without data loss
– Document the Serviceguard/HANA cluster configuration
Knowledge transfer
– Provide overview of HANA hardware, configuration, and networking
– Provide overview of HANA Studio and HANA administration
– Provide Overview of HANA backup and recovery

Solution validation and success criteria

This implementation progressed smoothly due to the thorough design that met the requirements, numerous evaluations during design reviews, a well-planned implementation, and the best experts to perform the installation.

SoH Testing

After the initial deployment of the SoH, development instance testing consisted of multiple sets of SoH application transactions to verify the results with the existing SAP ECC Production systems.

Once the Production Superdome X and Serviceguard configuration were complete the validation testing shifted toward the Serviceguard cluster. During initial Serviceguard testing, the new Serviceguard 12.1 SafeSync feature with SAP HANA was introduced.

The SafeSync feature provides both the automatic and the manual handling of HANA Serviceguard packages to maintain data integrity. In order to prevent data inconsistencies, the SafeSync mechanism tracks whether the secondary (failover) data is in sync. In the event of replication issues, SafeSync creates, blocks, and delays DB commits and prevents failover system from becoming the production instance. If HANA replication resynchronizes the failover system, the blocks are removed.

Serviceguard SafeSync provides the SoH workload higher system availability by allowing the Production SoH system to continue operation while synchronous replication cannot be achieved. Also, SafeSync achieves a higher level of data integrity because it prevented production use of secondary Failover data that is out of sync.

All subsequent Serviceguard failover testing with SafeSync configured and enabled resulted in zero transaction loss during all Serviceguard cluster test runs.

Deployment overview

The HANA TDI SoH deployment is a significant departure from the fully integrated HANA appliances. For this reason, the implementation must be painstakingly planned and executed and the users of the environment play a key role

The production TDI environment must be certified by SAP with defined Key Performance Indicators (KPIs). All aspects of CPU, Memory, and Disk Input/Output (I/O) performance must be tested to ensure that parameters are met by the deployment.

With the assistance of qualified HPE HANA consultants, the deployment can be greatly streamlined. In the case of this SoH workload, HPE HANA consultants were deployed to assist in the following areas:

- HPE Serviceguard for Linux cluster installation, configuration, and testing.
- SAP HANA replication configuration.

- SUSE Linux Enterprise Server (SLES) operating system installation to SAP HANA specifications
- HANA network configuration within Superdome X enclosure.

HPE Superdome Flex

The solution in this chapter used the Superdome platform. This section describes the Superdome Flex platform.

Superdome Flex combines the best of HPE's Superdome X and MC990 X technologies as shown in Figure 8-7.

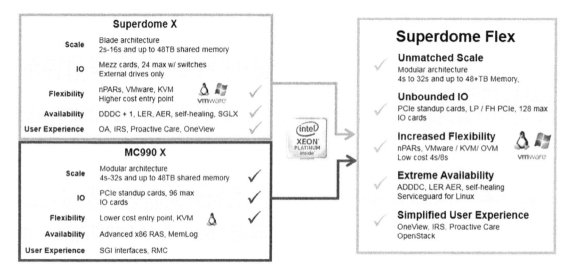

Figure 8-7 Superdome Flex Features

Here are some of the key points depicted in the diagram:

- In-memory design
- Scale of 4-32 sockets and 768 GB–48 TB memory capacity for scale-up workloads
- Expandable design
- Superdome Flex minimum memory configuration = 768 GB, maximum memory configuration = 48 TB, headroom for memory growth = 48 TB/768 GB = 62.5
- Low latency and high bandwidth fabric

Additional features include:

- A modular building-block design
- Up to eight four-socket modular building blocks
- Open management, open APIs with Redfish ecosystem, and support for OneView
- Hard partition support providing electrical isolation
- Reliability, Availability, and Scalability (RAS) features include the following:
- Error Analysis Engine that predicts hardware faults and initiates self-repair without operator assistance
- "Firmware First" approach, ensures error containment at the firmware level, including memory errors, before any interruption can occur at the OS layer

You may want to start your Data Management or Data Analytics initiatives with a smaller project, for example, start with one line of business or one application and then expand. You can incrementally scale as shown in Figure 8-8.

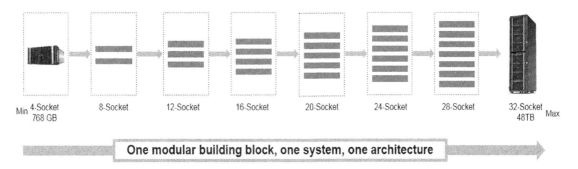

Figure 8-8 Superdome Flex incremental upgrades

This incremental scaling approach avoids overprovisioning with a large system. You can also upgrade without disruption using the modular building block approach. There are no "forklift" upgrades required with this design.

The next section shows an example of a Superdome Flex design. It does not go into the detail of the Superdome solution shown earlier in this chapter, but it shows the design at a high level.

Example Superdome flex implementation

This example is a Greenfield SAP S4 implementation in a phased multi-release approach over 2 years. The client's plan is to initially roll out S4 Finances and then Supply Chain and Order Management to follow. The initial sizing estimates from SAP for S4 Finance (S4FINs) is approximately 2TB of Memory and the sizing estimates of S4 Supply Chain (SC) and Order Management (OM) is approximately 4 TB.

CHAPTER 8
HPE HANA TDI Solution and Superdome Flex

The following are the business requirements of this solution:

- Provide a solution that supports initial rollout of S4 Finance.
- Scales to support S4 Supply Chain and Order Management rollout.
- Provides rapid business growth with a scalable HANA platform.
- Provide mission-critical platform to support the workload 24x7 worldwide operations.

The following are the corresponding technical requirements of the solution:

- Provide a scalable HANA solution capable of scaling to beyond four sockets and handling certified SAP HANA workloads to 8 TB and potentially beyond.
- Provide a HA clustering solution capable of providing local failover without any transaction loss in SAP HANA environment.
- Provide DRI solution for new S4 environment.

After considering the business and technical requirements for this solution, a Superdome Flex design was crafted. The Superdome Flex allows this HANA workload to scale by adding 4-Socket expansion blocks to initial base block. The result is a seamless and less disruptive growth path to manage growth for the HANA workload at minimal risk to the overall availability of the Business Critical HANA workloads in the user's environment (Figure 8-9).

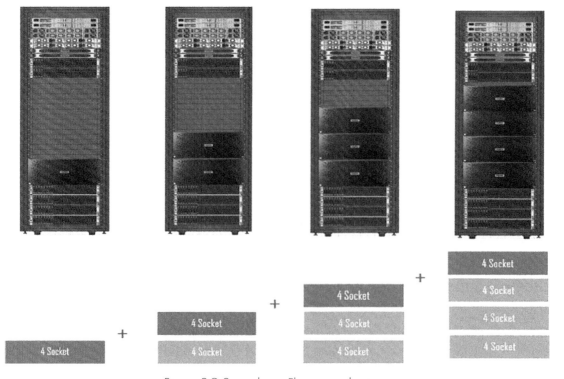

Figure 8-9 Superdome Flex upgrade steps

The client's production SAP S4/HANA architecture for the initial workload consisted of Qty (2) Superdome Flex Base servers. These servers consisted of both production and production HA Superdome Flex Base units. The Superdome Flex Base servers consisted of 4-socket Intel Skylake processor servers and 3 TB of Memory each. The S4 instance was also configured to into an HPE Serviceguard for Linux cluster with the production HA node. Serviceguard software facitltates auto-failover between the production S4 HANA database servers.

9 Memory-Driven Computing

INTRODUCTION

HPE is developing a radical new approach to future computing needs that will support the digital-centric world we are just beginning to experience and that will expand exponentially over the next few years. This new approach will result in a quantum leap in performance, enabling digitalization to realize its potential value.

The new approach began when Hewlett Packard Labs invested in a research project called "The Machine." The goal of "The Machine" project is to drive innovation around a concept called "Memory-Driven Computing" (MDC.) HPE believes that Memory-Driven Computing creates a new foundational architecture that will remove the limitations imposed by traditional computing and will create a new lifecycle of innovation, enabling solutions for challenges yet to come.

Before we discuss the technologies of MDC, it is important to understand the trends and associated challenges we face, given the traditional approach to computing that has defined the approach of the industry for the last several decades.

The evolution of IT consumption models

Digital disruption has become the latest catch-phrase being used by industry pundits and analysts, and as expected it has become viral. The disruption is real, but what is sometimes overlooked is that it has been triggered by the way in which a business uses technology to drive innovation around its operating models, customer engagement, and creation of new business models.

As digital technologies have evolved, businesses have looked to adapt ways to improve and become more competitive. Figure 9-1 illustrates how business has evolved to adapt new consumption models based on advances in digital technology. Note that the new models add to rather than replace existing models because each consumption model maintains its relevancy by continuing to deliver different value to the business.

CHAPTER 9
Memory-Driven Computing

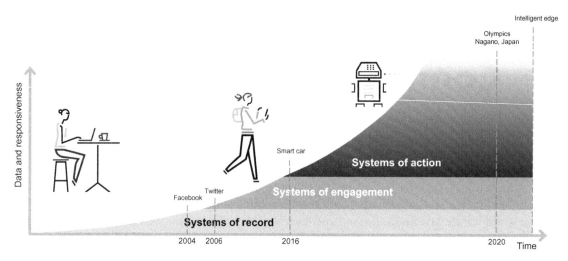

Figure 9-1 Evolution of IT consumption models

System of Record

Early adaption of information technology focused on what is known as "System of Record." In effect, this is simply recording what happened, generally mediated by people. In many cases, this is an electronic version of what was a paper record delivering accuracy and traceability. Companies adapted System of Record for enhancing productivity and lowering TCO. Examples would be transactions such as banking, retail, and so on.

System of Engagement

As the Internet has evolved and provided scale, both a broader market and highly personalized experience, the focus turned to a "System of Engagement" model. This supports highly interactive, near real-time applications that generally handle "messy" data such as social media. Systems of Engagement enable companies to create new and innovative ways to deliver their products and services along with the ability to collect and analyze customer data for use in marketing and the further creation of value that is relevant to their customers.

System of Action

We are now in the early stages of the "System of Action" model. This model supports a high-level of autonomy within the systems where machines either highly augment human decisions or fully control decisions and/or actions. A System of Action model requires processing in real time, is highly sensitive to latencies and jitter within the system, and needs high accuracy and traceability. Systems of Action works with both structured and unstructured data. Examples would be the Smart Grid and self-driving cars.

New processing paradigms

Among the fastest growing segments for compute is in the area of processors optimized for specific functions. The use of Graphic Processing Units (GPUs) for deep learning applications is likely the most prevalent in the trade press; however, applications involving virtual and augmented reality, predictive algorithms used in financial markets and genomic modeling are, among others, with specific processor requirements that are optimized when using nongeneral-purpose processors. Choices for architectures include Field Programmable Gate Array (FPGA,) GPU, Application Specific Integrated Circuit (ASIC,) System on a Chip (SoC,) Advanced RISC Machine (ARM,) and so on. Emerging architectures include neuromorphic, such as HPE Memristor, photonic, and quantum.

 Note

For an overview of HPE Moonshot and SoC, see the section at the end of this chapter.

Trends in application architecture

Software development has evolved over the last 20 years with the goal of achieving more agility without sacrificing quality. In addition to the movement toward DevOps and CI/CD, application architectures are shifting toward more modularity with the use of container technology and microservices. The benefits to this trend are substantial; however, it brings challenges such as latency incurred by switching between microservices and increased traffic flow due to data flowing between microservices.

The end of scaling as usual

In 1965, Gordon Moore wrote a paper in which he described an observation on the industry's ability to increase the density of transistors in integrated circuits. He projected that the density of transistors in integrated circuits would double each year, updated in 1975 to reflect a period of 2 years. The implication was that the industry would have a predictable way to increase performance while decreasing cost. Although this was an observation only, there are a number of physical laws that govern the physical aspects of the performance/cost curve.

The chart in Figure 9-2 shows the progression of the most significant physical properties that govern the performance of systems as they are currently architected. As you can see, we are nearing a plateau for some of these properties. This chart shows data through 2015 and current data supports this plateau.

CHAPTER 9
Memory-Driven Computing

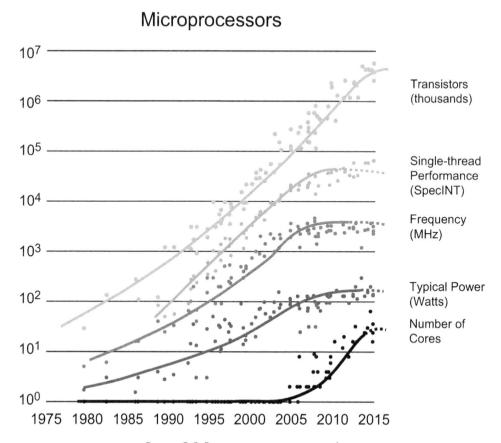

Figure 9-2 Properties governing scaling

Perhaps the most significant physical law is Dennard's Scaling. Dennard's Scaling is a scaling law that as transistors get smaller, their power density stays constant so that the power use stays in proportion with area. In the past, the effect from Dennard Scaling allowed chip manufacturers to increase clock frequencies with no impact to power consumption. As you can see on the chart, this broke down around 2006, effectively putting a halt to the ability of increasing clock frequencies to achieve higher performance. Since the frequency of the clock dictates the number of instructions a processor can execute per second, raw performance for single threaded code has plateaued as well.

Transistors have continued to shrink; however, since one processor could no longer run faster, successive generations of processors were manufactured with replicated copies of the processor core. This has benefited some workloads that are architected to take advantage of multiple cores; however, multicore processors require switching elements that increase the overall power consumption, effectively limiting the benefits of multiple cores.

The data revolution

As traditional processor scaling techniques are becoming ineffective, the timing could not be any worse as the volume, velocity, and variety of data ingested and consumed has skyrocketed. The chart in Figure 9-3 shows the growth of data created and its projected exponential rise. We have generated zettabytes of data that is nearly doubling every two years according to IDC. The exponential growth curve in rich data challenges conventional technology with its linear responses.

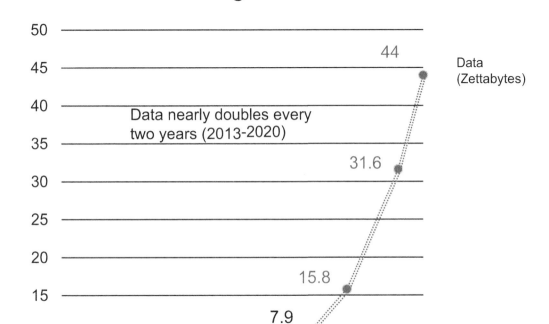

Figure 9-3 Skyrocketing data growth

In addition to its volume, the data consists of a heterogeneous mix of data types including text, graphics, and other multimedia with both structured and unstructured forms. Usage of the data can be both ephemeral and persistent, with some workloads using data in real time, while other workloads are leveraging historical data to create predictive models.

CHAPTER 9
Memory-Driven Computing

Even with advancements in big data frameworks, continuing the traditional storage hierarchy presents significant challenges for architecting data models to not only find answers to the questions we already know but also to use the data to discover the questions we should be asking.

Memory-driven computing (MDC)

At Hewlett Packard Enterprise we have been working on an innovative research program we call The Machine. HPE believes that the innovations created by The Machine research program will serve as the foundation for meeting current and future challenges and unlock the unrealized potential value of automation and digitalization en masse. The shift to Memory-Driven Computing is a necessary evolution to overcome the limitations of today's computing platforms and help power technology designed for a world of "intelligent everything."

Unlike conventional architectures that put the processor at the center of the computing paradigm where memory tends to be a scarce resource, the architecture of The Machine puts memory at the center. We call this new paradigm Memory-Driven Computing. Figure 9-4 depicts a comparison of today's Processor-Centric Computing versus Memory-Driven Computing.

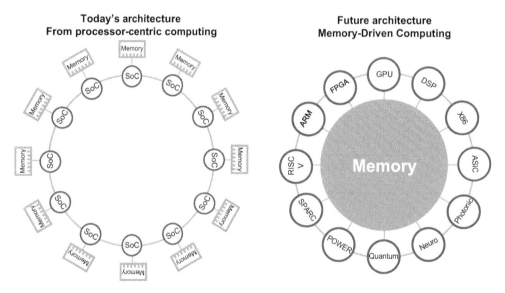

Figure 9-4 Processor centric versus memory driven

MDC takes advantage of emerging technologies, such as universal memory, photonics, and System-on-Chip (SoC) processors, to differentiate itself from conventional computing architectures. Conventional computing centers on processors. MDC shifts the emphasis to data in memory instead of bringing data to processors, MDC brings processing to data. MDC allows us to ingest, store, and manipulate massive data sets while simultaneously reducing energy/bit by orders of magnitude.

Key attributes

The key attributes of MDC appear in the following list:

- **Powerful**: The shift to MDC is a necessary evolution to overcome the limitations of today's computing platform, which will soon reach its physical and computational limits that is unable to optimize the unprecedented volume of data being created. MDC will serve as the foundation for all technology, underpinning everything from the data center to the Internet of Things to the next supercomputers.

- **Open**: HPE's goal is to make MDC an open ecosystem with many contributors. HPE believes the new architecture, combined with industry collaboration, can help solve the key challenges presented by today's data explosion. This open collaboration is what is behind the decision to put key development tools on GitHub and why the company saw it valuable to be a founding member of the Gen-Z Consortium.

- **Trusted**: Today, security is bolted on as an afterthought and the current design does not allow a lot of room for security. Modern attacks simply bypass antivirus and attack below the operating system where they are undetectable. MDC provides the opportunity to start over, to build in security from the start. HPE is focused on support below the operating system to provide isolation, detection of unknown compromises, and encryption of data in flight, in use and at rest.

- **Simple**: Current data models make it difficult to share and store data efficiently. MDC radically simplifies the way you use, store, and share data real time to manage your data efficiently. Benefit from a shorter path to persistence, rather than paying the costs of data translation as your data passes through various software layers.

- **Universal architecture**: HPE believes that MDC will serve as the foundation for all technology, underpinning everything from the data center to the Internet of Things to the next generation of supercomputers.

Core components

The MDC architecture is comprised of four core components:

- Shared Persistent Scale-Out Memory
- High-Performance Memory Fabric
- Task-Specific Processing
- Software Defined Composability

Figure 9-5 shows their inter-relationships, and the following sections describe the features and characteristics of each component in more detail.

CHAPTER 9
Memory-Driven Computing

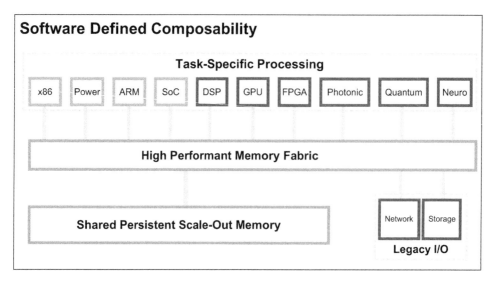

Figure 9-5 Components of MDC

Shared persistent scale-out memory

The key aspect of MDC is a virtually unlimited amount of shared nonvolatile memory. This provides an ability to collapse the typical storage hierarchy, limiting the need to copy data between memory and storage, which can be as much as 70% or 80% of processing time in the current architecture. The main features of the memory components are as follows:

- Memory is byte addressable with latency characteristics similar to DRAM.
- Memory is nonvolatile that means if there is a power loss, the memory maintains its contents.
- This memory is fast and operates at the speed of system memory.
- Minimal power is required to store vast amounts of data, which is less than the power consumed by a flash drive.
- Combining memory and storage in a stable environment to increase processing speed and improve energy efficiency.

High-performance memory fabric

MDC uses a photonics-based fabric that will provide a high-performance, low-latency way to connect pools of compute, memory, and external resources. HPE's MDC will use the Gen-Z open systems interconnect. Gen-Z is an open industry standard designed to provide memory semantic access to data and devices via direct-attached, switched, or fabric topologies. Key features and characteristics of this component include:

- The MDC Fabric enables future computing systems to communicate at the extreme scale and low-latency that data-centric computing will require to transition from today's storage and networking stacks to a radically simplified fabric-attached memory model with extreme energy-efficiency.

- HPE will leverage Gen-Z to extend the HPE composable framework to all components on the fabric.

- Photonics eliminates distance limitations and supports new innovative topologies.

- The use of Gen-Z enables the manipulation of huge data sets residing in large pools of fast, persistent memory.

- The Gen-Z protocol can address a flat fabric of 16 million components.

- Gen-Z could theoretically address 92 bytes, or 4096 yottabytes. Or a thousand times bigger than our digital universe today.

Task-specific processing

MDC enables a shift away from the dependence on general-purpose compute devices with support for a variety of compute architectures such as FPGA, GPU, ASIC, SoC, ARM, and in the future, Neuromorphic, Photonic, and Quantum.

- Applications will be able to achieve greater optimization by leveraging task-specific processing. Scale-out software architectures such as microservices can optimize performance by decomposing services based on code-specific compute requirements.

- This results in both higher compute performance and greatly reduced energy consumption.

Software-defined composability

MDC represents a fundamental shift in computing hardware architecture. As part of MDC, a complete re-evaluation of operating system principles, architecture, and services is required to enable systems that will operate at unprecedented scale. This will achieve radical simplify Data Center Automation and includes software to dynamically automate full composability of all resources (compute, fabric, and memory).

MDC software architecture

MDC required a fundamental re-thinking of computing hardware architecture. When looking at the various layers in the software architecture, we also went through a complete re-evaluation of operating system principles, architecture, and services to enable systems that will operate at unprecedented scale.

MDC software stack

HPE is developing system software technologies for a new class of applications that will allow for effective utilization of MDC's massive, distributed nonvolatile fabric-attached memory. Figure 9-6 shows a representation of the software stack for MDC.

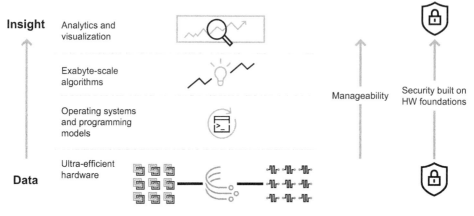

Figure 9-6 MDC software stack

The work in this area has two tracks: first, it will use familiar programming constructs to allow legacy applications to run with improved performance. This is referred to as Linux for The Machine/MDC. Linux for The Machine will keep the POSIX interface for backwards compatibility, but strip out unnecessary layers to take advantage of the new hardware and firmware.

MDC developer toolkit

As part of HPE's goal to make MDC an open ecosystem with many contributors, Hewlett Packard Labs has created a repository on GitHub (https://github.com/HewlettPackard/mdc-toolkit). This repository provides a unified entry point to several Hewlett Packard Labs tools and technologies that enable programmers to realize the benefits of MDC and persistent memory.

The MDC Developer Toolkit consists of a number of useful software system and application software that can be found at https://www.labs.hpe.com/the-machine/developer-toolkit and on the HPE GitHub site (https://github.com/HewlettPackard/mdc-toolkit). The sites contain example applications, programming and analytics tools, operating system support, and emulation/simulation tools.

Figure 9-7 shows an architectural representation of the MDC Developer Toolkit followed by a general description of the tools currently available as open source and optimized for MDC. HPE's objective in providing these tools is for learning and providing examples to early adaptors who are incubating MDC initiatives. HPE has no intention of offering these tools as supported products or services.

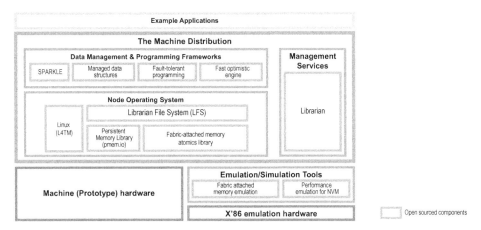

Figure 9-7 MDC Developer Toolkit

Sample applications

- *Image Search*: As part of The Machine program, we built a similarity search framework for high-dimensional objects. High-dimensional search enables important components of many analytic tasks in information retrieval such as document duplicate detection, machine learning such as nearest neighbor classification, and computer vision such as pose estimation and object recognition.

- *Large-Scale Graph Inference*: Graphical Inference is a remarkable algorithm for graph analytics that abstracts knowledge by combining probabilities and graph representations. It captures useful insights for solving problems such as malware detection, genomics analysis, IoT analytics, and online advertisement.

Data management and programming frameworks

- *Sparkle*. Optimized Spark to run 10x faster on large-memory systems. Hewlett Packard Labs has made changes available under the same license as Spark (the Apache 2.0 License), a fast and general cluster computing system for Big Data.

- *Multi-Process Garbage Collector*: Our state-of-the-art Multi-Process Garbage Collector (MPGC) delivers automatic memory management for applications, enabling multiple applications to share objects in memory, while avoiding memory management errors and memory leaks.

- *Managed Data Structures (MDS)*: A software library for persistent memory programming, which enables developers to declare familiar data structures such as lists and maps as persistent and reuse them across programs and languages, easing programming and data sharing at scale to reap the full benefits of MDC The Machine and other persistent memory architectures.

- *Fast optimistic engine for data unification*: A completely new database engine that speeds up applications by taking advantage of a large number of CPU cores and nonvolatile memory.

- *Fault-tolerant programming model for Non-Volatile Memory (NVM)*: Adapts existing multi-threaded code to store and use data directly in persistent memory. Provides simple and efficient fault-tolerance in the event of power failures or program crashes.

Persistent memory toolkit

- *Allocator layer for persistent shared memory:* Provides a collection of low-level abstraction layers that relieve the user from the details of mapping, addressing, and allocating persistent shared memory.

- *Application-transparent checkpoint with persistent memory:* System-level application-transparent tool for suspending and resuming Linux applications and Docker containers with persistent memory. It optimizes Checkpoint/Restore in user-space software tool to support fast checkpoint and restore with persistent memory, such as HPE Non-Volatile Dual Inline Memory Modules (NVDIMMs.)

- *Shoveller:* Shoveller is a scalable, memory-capacity efficient key-value store for very large-scale machines such as tens of terabytes or more memory and hundreds of CPU cores.

- *Non-Volatile Memory Manager (NVMM):* NVMM is a library written in C++ that provides simple abstractions for accessing and allocating NVM from fabric-attached Memory.

- *Radix Tree:* Radix Tree is a user-space library written in C++ that implements a radix tree that relies on fabric-attached memory atomics that are atomic primitives within a cache-incoherent memory-semantic fabric environment.

- *Write-Ahead-Logging Library (WAL) for Non-Volatile DIMMs:* WAL is the central component in various software that require atomicity and durability, such as Data Management Systems (DBMS).

Operating system support

- *Linux for MDC*: Hewlett Packard Labs is adapting the Linux Operating System to support MDC architectures such as The Machine project. Modifications to Linux include support for fabric-attached persistent memory and block device abstractions for persistent memory. Also included is an emulator that can be used by developers to explore the new APIs on industry standard machines.

Emulation and simulation tools

- *Fabric-Attached Memory Emulation*: An environment designed to allow users to explore the new architectural paradigm of The Machine, called Memory-Driven Computing. The emulation employs virtual machines performing the role of "nodes" in The Machine. Explore shared, global memory space, and expect it to behave like MDC on The Machine. Linux for Fabric-Attached Memory Emulation is also available to provide software for The Machine APIs and allow you to explore MDC using current hardware.

- *Performance emulation for NVM latency and bandwidth*: A DRAM-based performance emulation platform that leverages features available in commodity hardware to emulate different latency and bandwidth characteristics of future byte-addressable nonvolatile memory technologies.

Workload categories and examples

The composability that is inherent in MDC provides the resource capacity liquidity required to drive performance and enable new application architectures to produce outcomes with greater relevancy. Workload examples can range from IT Operations, Supply Chain Management, to applications analyzing economics and sociological trends.

MDC time machine

One of the most common examples of a workload category across all industries is the blending of historical and current data to provide a forecast of future intelligence. Interpretive algorithms and data are often very different, even within a single industry, but the use of time as the primary dimension of the data is perhaps the single common thread among all workloads in this category.

Workloads in this category will curate data, events, and outcomes in real time, providing the power of **Hindsight** to spot trends and anomalies within the data and find answers to questions that were not asked. Using current data, these workloads will use the knowledge gains via the process of hindsight and apply that knowledge to gain real-time **Insights** to optimize their decision process gaining more business agility. The leaders in any industry will look to gain **Foresight** by leveraging both hindsight and insight to predict variations of the future-to-further optimize their strategies (Figure 9-8).

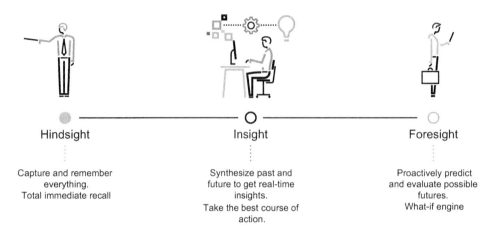

Figure 9-8 MDC Time Machine

In the next section, we will provide industry examples where the MDC Time Machine concept can be effectively utilized.

Transportation and logistics

Managing logistics arguably is the most computational complex challenges for the transportation industry. Everything from transporting packages to transporting people involve a large number of interconnected algorithms along with a multitude of variables, many of which typically exhibit sensitive dependence on the overall transportation ecosystem at any point in time.

Logistic systems have so many variables that conventional computing systems are not able to evaluate them at once. As a result, independent systems handle different aspects of the overall process and have local copies of relevant data, some of which are common to one or more other systems. An Enterprise Message Bus (EMS) is typically used for interconnecting systems.

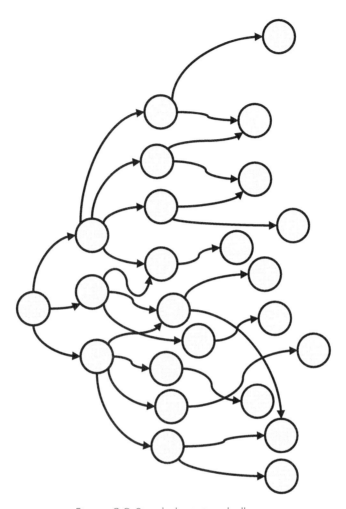

Figure 9-9 Simple logistics challenge

A simple example would be a delivery company that transports packages of variable size, shape, and weight from any location to any other location and assume within the United States. Figure 9-9 shows a simple diagram of a single location collecting packages to ship to multiple destinations. A single disruptive event such as severe weather in a major hub location or a mechanical breakdown could trigger a complete recalculation of the logistics models due to the sensitive dependency of key variables. This could potentially lead to all new route topologies. Even if an event is less impactful, recalculations of models are complex with many variables, leading from moderate to significant delays. As mentioned, the figure is a very simple diagram, with a realistic depiction being several orders of magnitude more complex.

Leveraging MDC, past data can be curated from real-life events to learn and understand historical root causes, patterns, and outcomes. Near real-time actionable intelligence that will trigger automatic route optimizations based on all factors related to the logistics models. In addition, models can be precomputed using various what-if scenarios so when an event occurs, a new and optimized route can be looked up in real time, avoiding costly delays.

Financial services

The financial industry heavily relies on simulations to value and analyze complex instruments and to manage risks. For financial portfolios with large numbers of assets or nonlinear instruments, such as derivatives, the simulations are very compute intensive. This forces financial institutions to make a trade-off between accuracy and speed. Financial institutions need, but cannot perform, accurate pricing and risk estimation in real time.

Leveraging MDC, application architects can explore solutions utilizing large-memory systems to accelerate simulations and reduce estimation time substantially. This provides firms with an ability to get accurate risk and price estimates in near real time instead of once or twice a day.

Leveraging the scale-out memory of MDC, architects can curate a broader and more granular set of data. This will enable them to quickly understand the historical and financial trend analysis to provide a financial story of a business and the types of decisions that impact their performance.

The compute elasticity and support for task-specific processor architectures of MDC will enable the capability and tools to price complex deals and to do a portfolio risk estimation in real time with a high accuracy level that will change the way investment decisions will be made.

Leveraging both scale-out memory and compute, fast calibration of complex financial models can be applied to model the market behavior for a more accurate projection of asset performance over time. This enables applications to be representative simulations that have been precomputed. In tests conducted by Hewlett Packard Labs, performance improvements up to 10,000 times have been demonstrated.

MDC Enterprise Service Bus (ESB)

Firms deploying Enterprise Messaging have used the same abstract architecture over two decades. This has led to a continuous churn of technology with an increased TCO with minimal quantifiable ROI. MDC for Enterprise Messaging leverages the scale-out compute and shared memory of MDC to deliver an architecture that radically simplifies trading systems and allows for frictionless scalability.

Figure 9-10 shows the abstract architecture of the MDC ESB.

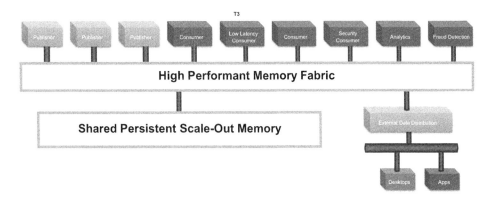

Figure 9-10 MDC Enterprise Service Bus

The architecture eliminates the Enterprise Message Bus and leverage the MDC Time Machine discussed in the previous section. Real-time updates from both external and internal publishers are captured and stored in shared, persistent memory. This would provide a common data fabric that can be accessed by any resources and support the various tiers of services required using standard Gen-Z primitives.

This architecture enables application developers to gain insights into what is happening in real time in the context of hindsight, and for some use cases, include foresight by leveraging precalculated "what-if" scenarios via simple look-up tables. Some potential use cases would include fraud detection, risk, and compliance and real-time balancing of managed funds.

Compute becomes a highly elastic resource where workloads and/or microservices can be composed/decomposed on the order of microseconds. This includes both persistent instantiations of a workload or a new instantiation to accommodate a subsecond burst.

MDC similarity search

As part of The Machine program, we built a similarity search framework for high-dimensional objects. High-dimensional search enables important components of many analytic tasks in information retrieval (for example, document duplicate detection), machine learning (for example, nearest neighbor classification), and computer vision (for example, pose estimation and object recognition). The search can be done over other data types such as images, documents, tweets, and time series. Image search is one example of an application that requires high-dimensional similarity search. Imagine if one has to search through all the recorded images to determine the presence of a person of interest specified by their picture. To accomplish such a task requires an accurate search on a large corpora of images that returns the images most similar to the query image. An enormous amount of computing power may be required to perform this search quickly and accurately.

CHAPTER 9
Memory-Driven Computing

This example features an image similarity search against 80 million images, where the user can choose an image, or take their own, and ask the system to return the five most visually similar images. We compared three architectural approaches to providing a solution:

- disk-based search with Hadoop (today's typical approach) (slow)
- in-memory search, optimized with indexed data (faster)
- simulated Machine, using NVM and can hold the entire index in memory, so the answers are already "there," we just need to look them up (fastest).

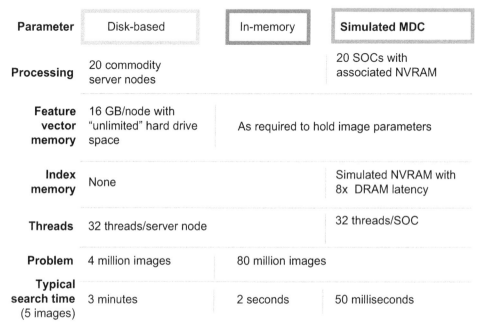

Figure 9-11 Similarity search configuration and results

Figure 9-11 shows the configuration for the three architectural approaches and the results achieved for each approach. One difference to highlight was that for the in-memory and simulated MDC, we increased the image databased by a factor of 20. Even with the increased sized of the database, we see a 60x performance gain using the MDC approach.

More details on this example, including the source code is available on the Hewlett Packard Enterprise GitHUB site (https://github.com/HewlettPackard/SimilaritySearch).

HPE Moonshot and System on a Chip (Soc)

One of the key components of memory-driven covered earlier is multiple SoC accessing the shared memory pool. Moonshot is the platform on which HPE developed the SoC concept. It is used for a variety of workloads, and different SoC are supported in the same enclosure. This is the basis of The Machine in which many different types of SoC may work on the same shared memory pool. In the case of Moonshot, workloads such as Hosted Desktop Infrastructure (HDI) have SoC dedicated to a high-performance desktop, such as a trader workstation.

Figure 9-12 shows the Moonshot solution and applications at the top, the internal network for connectivity, and then remote users and clients at the bottom.

Moonshot is a different form factor from other enclosures. Figure 9-12 shows the Moonshot chassis, cartridges on the left, and a diagram of the back of the chassis on the right:

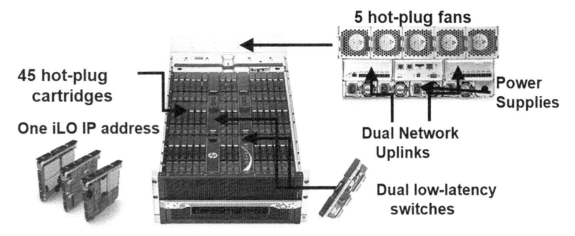

Figure 9-12 Moonshot Components

SoC is an example of HPE-developed technology that is used in production today and is one of the key building blocks of The Machine.

Summary

MDC is a powerful, open, trusted, simple, and new universal architecture that greatly surpasses the processing limitations of conventional computing. Simply stated, MDC brings processing to data memory instead of bringing data to processors. By taking advantage of emerging technologies (such as universal memory, photonics, and System-on-Chip (SoC) processors) and providing the resource capacity liquidity to drive performance, MDC unlocks the ability to ingest, store, and manipulate massive data sets while simultaneously reducing the energy/bit by orders of magnitude. These new capabilities will enable innovative application architectures and differentiated data-driven business models designed to take advantage of the tremendous opportunities that come with accelerating digital transformation.

Additional resources

The following URLs provide information on getting started with MDC.

- To get the latest updates on The Machine Project and Memory-Driven Computing, visit www.hpe.com/themachine

- Join The Machine User Group at https://www.labs.hpe.com/the-machine/user-group

    - For community discussions, sign up to our Slack group #themachineusergroup channel at https://www.labs.hpe.com/slack

    - Subscribe to "The Machine User Group" tab in the "Behind the scenes @ Labs" blog https://community.hpe.com/t5/Behind-the-scenes-Labs/bg-p/BehindthescenesatLabs/label-name/The%20Machine%20User%20Group#.WXZGN4jyscE. Register and click "subscribe to this label."

- Questions? Contact themachineusergroup@hpe.com

- Get access to the Memory-Driven Computing Developer Toolkit at https://www.labs.hpe.com/the-machine/developer-toolkit

- Follow us on our Hewlett Packard Labs social handles:

    - Twitter: @HPE_Labs

    - LinkedIn: "Hewett Packard Labs"

    - "Hewlett Packard Enterprise" YouTube page—The Machine and Hewlett Packard Labs channels

    - Instagram: HPE

    - Facebook: Hewlett Packard Enterprise

10 Worldwide VDI Deployment on HPE SimpliVity

INTRODUCTION

The practice of running a desktop operating system on a virtual machine (VM) hosted on a server, known as Virtual Desktop Infrastructure (VDI), has now become mainstream. It can be implemented in a variety of ways and on several different platforms. This chapter covers a recommended VDI deployment on HPE SimpliVity.

Solution requirements

The following are the drawbacks related to a traditional desktop environment:

- With growth of the global business, there is an urgent need to simplify, standardize, and protect access to corporate environment for developers, contractors, traveling managers, and regular users.
- Lifecycle management of tens of thousands of corporate desktops and laptops worldwide is too slow, cumbersome, unreliable, and expensive.
- Regulatory compliance is difficult to check and audits consume a lot of time.
- High Availability (HA) and Disaster Recovery (DR) requirements are not met when users own endpoints (laptops), resulting in excessive buildout of backup and restore infrastructure in each geographic region.

To overcome these drawbacks, the success criteria for a new desktop architecture are as follows:

- Faster response time
- Retain end user application performance at current levels
- Improve restore times in case of a desktop or application failure
- Deliver five nines availability and 100% DR capability
- Simplify HA and DR infrastructure and standardize on recovery methods across the globe
- Speed up deployment of new users
- Reduce lifecycle cost per user

The specific environment covered in this solution exhibited the following high-level characteristics:

- Tens of thousands of users in the Americas, EMEA, and APJ
- A mix of persistent and nonpersistent VDI users
- Persistent: 4 vCPUs, 12 GB RAM, 30 IOPs, 100 GB storage, 6:1 overcommit
- Non-persistent: 2 vCPU, 8 GB RAM, 20 IOPs, 75 GB storage, 8:1 overcommit

Based on all of these considerations, and more, the following solution was crafted for a global, uniform VDI solution based on SimpliVity Gen10 architecture.

- Citrix XenDesktop on VMware vSphere VDI solution deployed on SimpliVity Gen10 nodes
- HA clustering (PODs) of SimpliVity nodes in area data centers
- Active/Active implementation and full DR fail over between pairs of data centers within a geographic area
- Area-based lifecycle management with global monitoring and reporting.
- HPE Pointnext services for initial implementation, ongoing support, and lifecycle maintenance
- HPE Financial Services for CAPEX to OPEX conversion and Flex Capacity options

The following section describes the solution overview.

Solution overview

HPE SimpliVity is a hyperconverged solution for client virtualization and VDI, delivering on-target user performance, cost-effective economics, and enterprise-level data protection and resiliency. Compute and storage optimizations are achieved by easily combining storage and compute nodes, highly efficient data virtualization, and aggressive inline deduplication and compression that dramatically reduce raw storage capacity requirement. Scalability is achieved by clustering and subsequent federation of SimpliVity nodes, while retaining global unified management.

In each geographic area (USA, EMEA, APJ), the proposal includes a geographical disbursed two-site identical sets of VDI clusters configured in an Active/Active configuration to ensure high availability, disaster recovery and similar access latency for assigned users.

In each data center, the VDI infrastructure was, first, sized up relative to the target number of assigned active users. Then the infrastructure was doubled to address DR requirement.

Data center sizing relied on recommendations available on SimpliVity Login VSI numbers published by Citrix and HPE (https://h20195.www2.hpe.com/v2/Getdocument.aspx?docname=a00017191enw) and further extrapolated upon leveraging Intel's Xeon Scalable Processors performance estimates (https://www.intel.com/content/www/us/en/benchmarks/xeon-scalable-benchmark.html)

From a physical infrastructure perspective, the nodes in each data center are arranged in PODs. All of the nodes are the same size and, depending on the number of nodes in a POD, the PODs can be full or half. Data protection is provided by RAID (local protection) + RAIN (across two nodes) for both POD types. Figure 10-1 depicts a full and half POD and the number of users that can be supported for this specific solution (the users supported is not a general number, it is dependent on the user mix and other factors):

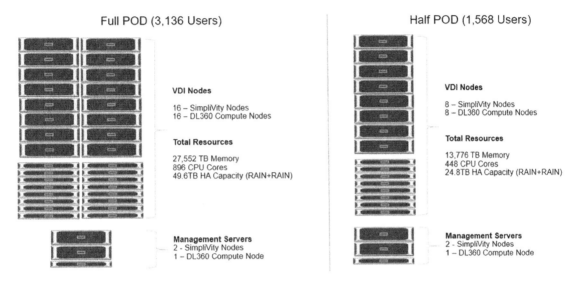

Figure 10-1 VDI SimpliVity PODs

Two types of building blocks are needed in order to deploy this Citrix VDI infrastructure:

- VDI Block—One DL380Gen10 SimpliVity node for compute and storage plus one DL360Gen10 node for additional compute. These blocks host VMs and user desktops.

- Management Block—Two DL380Gen10 SimpliVity nodes for compute and storage plus one DL360Gen10 node for additional compute. These blocks host Citrix logic (PVS, SQL, License Server, StoreFront, and so on.)

The eight SimpliVity nodes comprise a VMware cluster and compute nodes are added to it in the Full POD in the diagram.

The infrastructure shown in Figure 10-1 will support up to the required number of VDI users worlwide. Note that while there are 8 or 16 VDI Blocks in a POD or a Half-POD, respectively, only one Management Block is required per POD or Half-POD. Dividing the number of users per region by the target number of users per POD, 3136 for a Full POD and 1568 for a half POD would provide the scale-out quantity for this client by region.

CHAPTER 10
Worldwide VDI Deployment on HPE SimpliVity

Solution details

This section outlines the key components that constitute the VDI Block and Management Block.

VDI Block (1x SimpliVity DL380 Gen10 and 1x DL360 Gen10 ProLiant servers)

Figure 10-2 shows the DL380 Gen10 SimpliVity node for compute and storage.

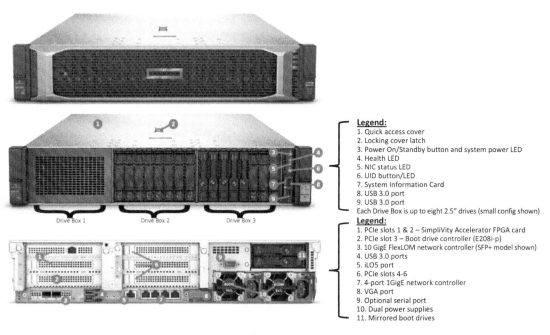

Figure 10-2 SimpliVity DL380 Gen10 "Small" Node

Figure 10-3 shows the DL360 Gen10 node for additional compute.

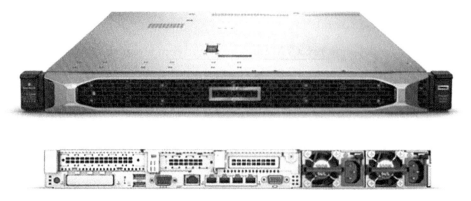

Figure 10-3 DL360 Gen10 Node

Table 10-1 shows the VDI Block Bill of Materials (BOM).

Table 10-1 VDI Block BOM

Qty	Product #	Product description
1	Q8D81A	HPE SimpliVity 380 Gen10 Node
1	Q8D81A 001	HPE SimpliVity 380 Gen10 VMware Solution
1	826856-L21	HPE DL380 Gen10 Intel Xeon-Gold 5120 (2.2 GHz/14-core/105 W) FIO Processor Kit
1	826856-B21	HPE DL380 Gen10 Intel Xeon-Gold 5120 (2.2 GHz/14-core/105 W) Processor Kit
1	826856-B21 0D1	Factory integrated
2	Q8D87A	HPE SimpliVity 384G 12 DIMM FIO Kit
1	Q8D91A	HPE SimpliVity 380 for 4000 Series Small Storage Kit
1	875241-B21	HPE 96W Smart Storage Battery (up to 20 devices/145 mm Cable) Kit
1	875241-B21 0D1	Factory integrated
1	804331-B21	HPE Smart Array P408i-a SR Gen10 (8 Internal Lanes/2 GB Cache) 12 G SAS Modular Controller
1	804331-B21 0D1	Factory integrated
1	700751-B21	HPE FlexFabric 10 Gb 2-port 534FLR-SFP+ Adapter
1	700751-B21 0D1	Factory integrated
2	830272-B21	HPE 1600W Flex Slot Platinum Hot Plug Low Halogen Power Supply Kit
2	830272-B21 0D1	Factory integrated
1	BD505A	HPE iLO Advanced including 3yr 24x7 Tech Support and Updates 1-server LTU
1	BD505A 0D1	Factory integrated
1	Q8A60A	HPE OmniStack 8-14c 2P Small SW
1	733664-B21	HP 2U Cable Management Arm for Easy Install Rail Kit
1	733664-B21 0D1	Factory integrated
1	733660-B21	HP 2U Small Form Factor Easy Install Rail Kit
1	733660-B21 0D1	Factory integrated
1	826706-B21	HPE DL380 Gen10 High Performance Heat Sink Kit
1	826706-B21 0D1	Factory integrated
1	867959-B21	HPE ProLiant DL360 Gen10 8SFF Configure-to-order Server
1	867959-B21 ABA	HPE DL360 Gen10 8SFF CTO Server
1	860665-L21	HPE DL360 Gen10 Intel Xeon-Gold 5120 (2.2 GHz/14-core/105 W) FIO Processor Kit

(Continued)

Table 10-1 VDI Block BOM—cont'd

Qty	Product #	Product description
1	860665-B21	HPE DL360 Gen10 Intel Xeon-Gold 5120 (2.2 GHz/14-core/105 W) Processor Kit
1	860665-B21 0D1	Factory integrated
16	815101-B21	HPE 64GB (1x64GB) Quad Rank x4 DDR4-2666 CAS-19-19-19 Load Reduced Smart Memory Kit
16	815101-B21 0D1	Factory integrated
2	868818-B21	HPE 480 GB SATA 6 G Read Intensive SFF (2.5 inch) SC 3 yr Wty Digitally Signed Firmware SSD
2	868818-B21 0D1	Factory integrated
1	804326-B21	HPE Smart Array E208i-a SR Gen10 (8 Internal Lanes/No Cache) 12 G SAS Modular Controller
1	804326-B21 0D1	Factory integrated
1	700751-B21	HPE FlexFabric 10 Gb 2-port 534FLR-SFP+ Adapter
1	700751-B21 0D1	Factory integrated
2	455883-B21	HPE BladeSystem c-Class 10 Gb SFP+ SR Transceiver
2	455883-B21 0D1	Factory integrated
1	339778-B21	HP RAID 1 Drive 1 FIO Setting
2	830272-B21	HPE 1600W Flex Slot Platinum Hot Plug Low Halogen Power Supply Kit
2	830272-B21 0D1	Factory integrated
1	BD505A	HPE iLO Advanced including 3 yr 24x7 Tech Support and Updates 1-server LTU
1	BD505A 0D1	Factory integrated
1	734811-B21	HPE 1U Cable Management Arm for Rail Kit
1	734811-B21 0D1	Factory integrated
1	874543-B21	HPE 1U Gen10 SFF Easy Install Rail Kit
1	874543-B21 0D1	Factory integrated
	HA113A1	HPE Installation SVC
1	HA113A1 5A0	HPE Entry 300 Series Install Service
2	455883-B21	HPE BladeSystem c-Class 10 Gb SFP+ SR Transceiver
	H7J35A3	HPE 3Y Foundation Care 24x7 wDMR SVC
1	H7J35A3 WAG	HPE DL360 Gen10 Support
1	H7J35A3 Z9X	HPE SVT 380 Gen10 Node (1 Node) Support
1	H7J35A3 ZA4	HPE OmniStack 8-14c 2P Small Support
	HA114A1	HPE Installation and Startup SVC

(Continued)

Table 10-1 VDI Block BOM—cont'd

Qty	Product #		Product description
1	HA114A1	5LY	HPE SimpliVity 380 HW Startup SVC
	HA124A1		HPE Technical Installation Startup SVC
1	HA124A1	5LZ	HPE SimpliVity 380 SW Startup SVC

Management block (2x SimpliVity DL380 Gen10 and 1x DL360 Gen10 ProLiant servers)

For the management block, the only changes compared to the VDI block are to the processor power and memory footprint.

Table 10-2 shows the Management block Bill of Materials (BOM).

Table 10-2 Management block BOM

Qty	Product #	Product description
2	Q8D81A	HPE SimpliVity 380 Gen10 Node
2	Q8D81A 001	HPE SimpliVity 380 Gen10 VMware Solution
2	826866-L21	HPE DL380 Gen10 Intel Xeon-Gold 6130 (2.1 GHz/16-core/120 W) FIO Processor Kit
2	826866-B21	HPE DL380 Gen10 Intel Xeon-Gold 6130 (2.1 GHz/16-core/120 W) Processor Kit
2	826866-B21 0D1	Factory integrated
4	Q8D85A	HPE SimpliVity 240G 12 DIMM FIO Kit
2	Q8D91A	HPE SimpliVity 380 for 4000 Series Small Storage Kit
2	875241-B21	HPE 96 W Smart Storage Battery (up to 20 devices/145 mm cable) Kit
2	875241-B21 0D1	Factory integrated
2	804331-B21	HPE Smart Array P408i-a SR Gen10 (8 Internal Lanes/2 GB Cache) 12 G SAS Modular Controller
2	804331-B21 0D1	Factory integrated
2	700751-B21	HPE FlexFabric 10 Gb 2-port 534FLR-SFP+ Adapter
2	700751-B21 0D1	Factory integrated
4	830272-B21	HPE 1600 W Flex Slot Platinum Hot Plug Low Halogen Power Supply Kit
4	830272-B21 0D1	Factory integrated
2	BD505A	HPE iLO Advanced including 3 yr 24x7 Tech Support and Updates 1-server LTU
2	BD505A 0D1	Factory integrated

(Continued)

Table 10-2 Management block BOM—cont'd

Qty	Product #	Product description
2	Q8A68A	HPE OmniStack 16-22c 2P Small SW
2	826706-B21	HPE DL380 Gen10 High Performance Heat Sink Kit
2	826706-B21 0D1	Factory integrated
2	733664-B21	HP 2U Cable Management Arm for Easy Install Rail Kit
2	733664-B21 0D1	Factory integrated
2	733660-B21	HP 2U Small Form Factor Easy Install Rail Kit
2	733660-B21 0D1	Factory integrated
1	867959-B21	HPE ProLiant DL360 Gen10 8SFF Configure-to-order Server
1	867959-B21 ABA	HPE DL360 Gen10 8SFF CTO Server
1	860687-L21	HPE DL360 Gen10 Intel Xeon-Gold 6130 (2.1 GHz/16-core/125 W) FIO Processor Kit
1	860687-B21	HPE DL360 Gen10 Intel Xeon-Gold 6130 (2.1 GHz/16-core/125 W) Processor Kit
1	860687-B21 0D1	Factory integrated
12	815100-B21	HPE 32 GB (1x32 GB) Dual Rank x4 DDR4-2666 CAS-19-19-19 Registered Smart Memory Kit
12	815100-B21 0D1	Factory integrated
2	868818-B21	HPE 480GB SATA 6G Read Intensive SFF (2.5 inch) SC 3yr Wty Digitally Signed Firmware SSD
2	868818-B21 0D1	Factory integrated
1	804326-B21	HPE Smart Array E208i-a SR Gen10 (8 Internal Lanes/No Cache) 12G SAS Modular Controller
1	804326-B21 0D1	Factory integrated
1	700751-B21	HPE FlexFabric 10 Gb 2-port 534FLR-SFP+ Adapter
1	700751-B21 0D1	Factory integrated
2	455883-B21	HPE BladeSystem c-Class 10 Gb SFP+ SR Transceiver
2	455883-B21 0D1	Factory integrated
1	339778-B21	HP RAID 1 Drive 1 FIO Setting
2	830272-B21	HPE 1600 W Flex Slot Platinum Hot Plug Low Halogen Power Supply Kit
2	830272-B21 0D1	Factory integrated
1	BD505A	HPE iLO Advanced including 3 yr 24x7 Tech Support and Updates 1-server LTU
1	BD505A 0D1	Factory integrated
1	734811-B21	HPE 1U Cable Management Arm for Rail Kit

(Continued)

Table 10-2 Management block BOM—cont'd

Qty	Product #	Product description
1	734811-B21 0D1	Factory integrated
1	874543-B21	HPE 1U Gen10 SFF Easy Install Rail Kit
1	874543-B21 0D1	Factory integrated
	HA113A1	HPE Installation SVC
1	HA113A1 5A0	HPE Entry 300 Series Install Service
4	455883-B21	HPE BladeSystem c-Class 10 Gb SFP+ SR Transceiver
	H7J35A3	HPE 3Y Foundation Care 24x7 wDMR SVC
1	H7J35A3 WAG	HPE DL360 Gen10 Support
2	H7J35A3 Z9X	HPE SVT 380 Gen10 Node (1 Node) Support
2	H7J35A3 ZC0	HPE OmniStack 16-22c 2P Small Support
	HA114A1	HPE Installation and Startup SVC
2	HA114A1 5LY	HPE SimpliVity 380 HW Startup SVC
	HA124A1	HPE Technical Installation Startup SVC
2	HA124A1 5LZ	HPE SimpliVity 380 SW Startup SVC

Rack Power requirements

At this scale, a POD becomes a high-density installation, which puts specific requirements on power and cooling infrastructure in a data center. Using HPE Power Advisor, we estimate the maximum compute and storage load per rack to be about 8.5KVA and max BTU reaching slightly over 28,500 BTU/hr.

CHAPTER 10
Worldwide VDI Deployment on HPE SimpliVity

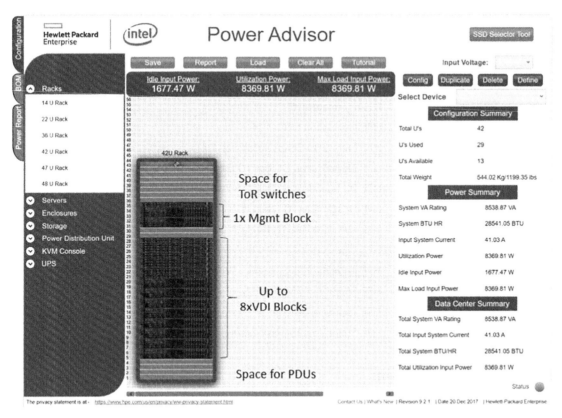

Figure 10-4 Rack Power and Cooling

In addition to the compute and storage equipment, the racks will contain redundant Top-of-Rack (ToR) switches, transceivers and cables, PDUs and other relevant components. The recommendation is that rack space be configured for 9.5KVA and 32,000 BTU/hr of redundant power and cooling.

Solution deployment

To facilitate design and deployment, the HPE Pointnext Services organization was engaged from the start of the project. Given project complexity and geographic reach, the following services were recommended for this solution:

- Building on-premises pilot infrastructure to verify performance and availability of the proposal. This also included development and testing of customer-specific scripts for automated deployment and integration in the existing data center management framework.

- Development of the overall deployment plan and management geography-specific implementation efforts, utilizing HPE Pointnext personnel as well as local system integrators.

- Working with HPE Factory Express to build, integrate, and ship complete racks of equipment that include HPE and third-party components (customer-supplied racks, PDUs, ToR switches, in-rack cabling, labeling, and so on).

- Managing local installation, setup, on-site configuration, and integration of the new equipment in the data center.

- Assisting the customer with VMware and Citrix expertise.

- Knowledge transfer for lifecycle management of the environment.

- Manage the entire environment as a worldwide service.

- Deploy the environment using a consumption model arrangement.

Summary

The solution covered in this chapter addresses the initial problem of a worldwide VDI solution. As with any solution, there were many ways to solve this problem, but SimpliVity was identified as the ideal solution. We were able to craft the PODs in such ways that they were easily expanded with additional computer nodes. The RAID and RAIN data protection methods were ideally suited for this solution as well. The software used to implement VDI, the implementation plan, support plan, and consumption-based services model combined to make this a complete solution. The use of consumption-based services are discussed in Chapter 15. These provided a significant basis for the procurement of this solution that was deployed over a 12-month period and ensured that no overprovisioning was required while it was being rolled out.

11 OneView Integration

INTRODUCTION

HPE OneView is HPE's solution for providing physical infrastructure provisioning across a large portion of HPE's enterprise portfolio. It was first introduced as, primarily, a web-based management tool that an IT Administrator could use to quickly configure and provision HPE's enterprise infrastructure. The web interface was the main initial focus, even though there has always been a very powerful Representational State Transfer (REST) Application Programming Interface (API), which could be written to in a number of ways, using scripting languages such as PowerShell or Python.

Nowadays, however, the emphasis on how to use OneView revolves around the use of RESTful APIs because these have become such a core part of the power of OneView and its ability to integrate with a much larger ecosystem of tools and solutions. Figure 11-1 illustrates many of the current integrations and the list is growing daily.

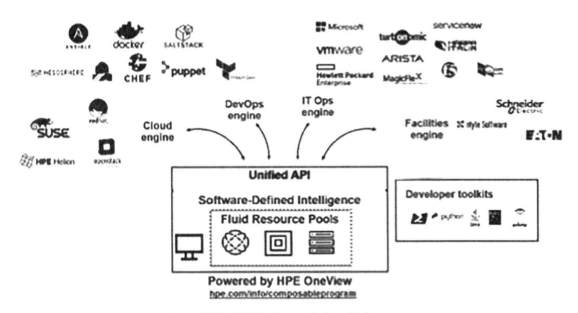

Figure 11-1 HPE Software Defined Infrastructure

RESTful-based industry solutions

Given its powerful RESTful API capabilities, HPE OneView is ideally purposed to ride the next wave of innovations. Businesses seeking a competitive edge are making large investments in IT resources that make the best use of industry leading tools and technologies and have become increasingly well-versed in how to integrate them into their IT processes and procedures. REST has emerged as the architectural style of choice for lightweight, networked-based application development. REST can be used from a variety of tools and almost any programming language. HPE OneView has an extensive and flexible RESTful-based Application Programming Interface (API), making it extremely powerful whether it be for coding custom integrations or when integrating with other RESTful-based industry solutions. HPE OneView's published API facilitates both scenarios extremely well.

HPE OneView can thus be seen as an "enabler" that allows organizations to leverage their existing IT investments. To cite just one example, HPE OneView's core capabilities become extremely powerful when paired with HPE Synergy (see the HPE Synergy chapter) as the underlying infrastructure to deliver composable infrastructure. This chapter will explore more of OneView's capabilities and the opportunities they provide for rapid innovation and advancements.

How HPE OneView works

The core functionality of HPE OneView revolves around a template-driven automation engine approach. The usage of templates and profiles is designed to interact with and represent the physical hardware infrastructure, but more in terms of a software construct. Server, storage, and networking are all addressable by these templates and profiles. It fosters collaboration between these individual departments within an organization to help enforce the best practices that they have developed.

HPE Image Streamer

With the addition of Image Streamer, HPE OneView can address the layer above the hardware, into the operating environment layer. This capability creates an efficiency and agility that can be used by the broader ecosystem of enterprise toolsets and solutions. As shown in Figure 11-1, there is extensive integration that is a part of HPE OneView. Many of them are available on Hewlett Packard Enterprises GitHub page listed at: https://github.com/HewlettPackard

A single unified workflow

So what does this mean in terms of real-world usage? In the traditional sense, the hardware and software have requirements that need to be addressed.

From a hardware infrastructure perspective, the following needs to be completed:

- Servers need to be configured to meet the application needs
- Storage needs to be zoned, capacity needs to be allocated, and ultimately assigned to the server or servers
- The network or networks need to be addressed as far as tagged or untagged traffic, Virtual Local Area Network (VLAN) assignments, bandwidth allocation, and so on

Once the hardware is addressed, the software layer comes next. Requirements for software include:

- The operating environment needs to be configured, patched, and hardened
- The middleware, including development tools, need to be provisioned
- The application needs to be deployed, configured, patched, hardened, and so on

The above steps, although oversimplified for the purposes of this example, usually represent multiple teams within an organization as well as multiple process, procedures, and tools. In a majority of instances, many of the same tools referenced above are being utilized. With HPE OneView, these steps can be addressed as part of a single workflow, and from a single tool, from a single team. This does not negate the need for, or the work that these teams perform, but rather uses and enforces the work done from a single controllable workflow or process.

Example of HPE OneView integration with Ansible

One example of the type of integration outlined earlier is what HPE has done with Ansible. Ansible performs automated software provisioning, configuration management, and application deployment. It deals with all of this functionality from the operating system layer and above and does so through the use of playbooks.

By using HPE OneView's integration with Ansible, and by referencing many of the sample playbooks created by HPE, an organization can quickly and easily add the required tasks sections to their existing playbook(s). Figure 11-2 shows how it is possible to add the physical hardware provisioning and configurations steps into any existing workflows and processes.

CHAPTER 11
OneView Integration

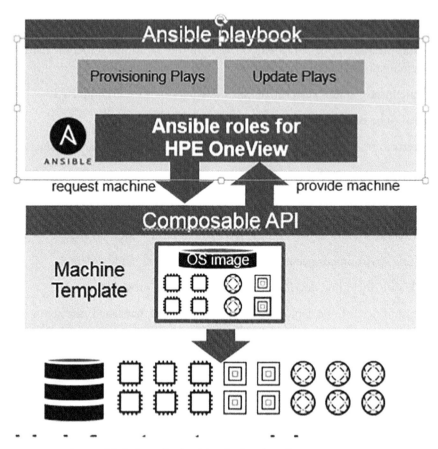

Figure 11-2 Sample Ansible Playbook with HPE OneView

Figure 11-3 illustrates how Ansible Playbook code can be used to create a server profile from an existing Server Profile Template within HPE OneView. The example also specifies certain specific fields and the relevant variables highlighted below:

- The Server Profile Template to leverage for the Server Profile creation
- Enclosure Group
- Hardware Type
- Network Name

```
- hosts: all
  vars:
    - config: "{{ playbook_dir }}/oneview_config.json"
    # Set the name of the server profile template that will be used to provision the server profile
    - ov_template: SY 480 Gen9 1
    - ov_template: spt1
    # Set the name of an existing enclosure group to run this example
    - enclosure_group_name: eg
    # Set the name of an existing server hardware type to run this example
    - server_hardware_type_name: SY 480 Gen9 2
    # Set the name of an existing ethernet network
    - network_name: eth1
  tasks:
    - name: Create a Server Profile from a Server Profile Template
      oneview_server_profile:
        config: "{{ config }}"
        data:
          serverProfileTemplateName: "{{ ov_template }}"
          name: "{{ inventory_hostname }}"
      delegate_to: localhost
      register: result

    - debug: msg="{{ result.msg }}"

    - debug: var=server_profile
    - debug: var=serial_number
    - debug: var=server_hardware
    - debug: var=compliance_preview
    - debug: var=created

    - name: Create a Server Profile with connections
      oneview_server_profile:
        config: "{{ config }}"
        data:
          name: "{{ inventory_hostname }}-with-connections"
          connections:
            - id: 1
              name: connection1
              functionType: Ethernet
              portId: Auto
              requestedMbps: 2500
```

Figure 11-3 Sample Ansible Playbook Code

The task that executes is the creation of the Server Profile based on the information provided. The full playbook can be found here: https://github.com/HewlettPackard/oneview-ansible/blob/master/examples/oneview_server_profile.yml

Because Ansible is being used to handle the physical infrastructure layer, an organization can take advantage of the platform much quicker with a faster time to value associated with an HPE OneView/Synergy solution.

Example of HPE OneView used with Docker

Another popular use case that has grown in interest exponentially is the usage of Docker Containers. Docker Containers are a key component in modernizing IT and application development within an organization. Docker is also a perfect workload to run on HPE Synergy and leverage the Composability and Automation that HPE OneView, HPE Image Streamer and Docker can provide.

Figure 11-4 represents a Reference Configuration for Docker Enterprise Edition (EE) Standard on HPE Synergy with HPE Synergy Image Streamer. The full paper can be found here: https://h20195. www2.hpe.com/v2/Getdocument.aspx?docname=a00008645enw

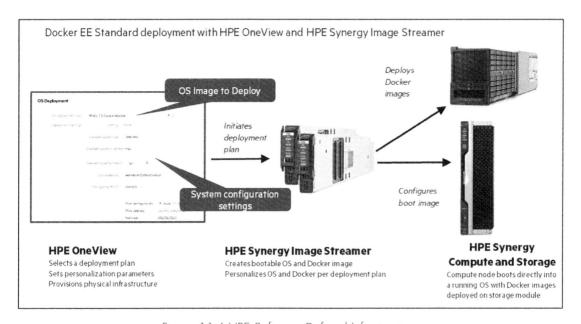

Figure 11-4 HPE Software Defined Infrastructure

The Docker integration that is being leveraged for this reference configuration represents some detailed interoperability between Docker, HPE Synergy Composer (OneView), HPE Image Streamer as well as VMware ESXi. ESXi is hosting the Docker Management plane consisting of the Universal Control Plane and Docker Trusted Registry as virtual machines, while the Docker worker nodes will be deployed on RedHat. Both Docker and ESXi are being provisioned via Image Streamer deployment plans and golden images. The Docker deployment plan associated with the Docker worker node provisioning is also running plan scripts. These scripts customize the environment and add the newly deployed, bare metal Docker worker nodes to the Docker Swarm Cluster. The automation and integration contained as part of this reference configuration is responsible for the end-to-end deployment of the entire solution including the physical infrastructure, Operating environment and application.

Summary

HPE OneView is an integrated management appliance that transforms HPE compute, storage, and networking into intelligent, software-defined infrastructure. This statement succinctly describes HPE OneView's mission statement and role in HPE's solutions portfolio. OneView is an enabler of automation and efficiency that drives tremendous value within an IT organization. When coupled with the partners associated with HPE's Composable Infrastructure ecosystem, it takes on an even more valuable role. In this chapter, we highlighted two of those partners specifically, Ansible and Docker, but there is a whole host of these ecosystem partners listed on HPE's Developers Hub and the list is growing daily. The link to the Developer's hub can be found here: https://hpe.com/info/composableprogram

The story does not end there since end users are limited only by their imagination. HPE OneView's RESTful API and multiple scripting language support allows customers to take HPE OneView to new heights every day. For more information on how HPE OneView can provide the foundation for your organization's software-defined infrastructure, visit: https://www.hpe.com/us/en/integrated-systems/software.html

12 High Performance Computing

INTRODUCTION

This chapter covers the creating of a networked compute infrastructure to run massively parallel scientific and deep learning programs. Massively parallel compute problems require architectural solutions that allow different parts of a scientific program to be run simultaneously on many individual servers in a coordinated manner. Since these programs often exchange information while running, a high-performance network is also required. Deep learning programs require specialized hardware within specially designed servers that have multiple Graphics Processing Unit (GPU) resources.

There were many key constraints and considerations when creating the design, including:

- **High-processing performance**—These scientific problems are compute-intensive and usually require servers with many cores and high frequencies. Some compute problems, and especially deep learning problems, can benefit from additional memory, and most can benefit from GPU acceleration. A balance of servers that have a modest amount of memory, a large amount of memory, and GPU acceleration is needed.

- **High-network performance**—Parallel processing programs exchange messages between servers to communicate and synchronize their work. Since work on an individual compute server could pause waiting for a message, high-network performance is required. Deep learning systems can generate massive amounts of data that need to be collect and stored. A high-performance network is essential to evacuate the data to a safe repository.

- **Reliable management nodes**—High-performance compute Infrastructures require a separate set of nodes that provide workload distribution. These nodes must be reliable and redundant to provide fast and efficient tasks to the high-performance cluster.

- **Cost**—High-performance compute solution sizing is different than standard business solutions in that the cost can be capped by the size of the grant if indeed the installation is in an educational institution which is the case in this example. A design goal is to provide the maximum amount of compute infrastructure that can be purchased under this cap within the bounds of the datacenter's space, power, and cooling.

- **Floor space**—Floor space is limited in this environment, and as such, the solution needs to be as dense as possible.

- **Power and cooling**—Once the density of the solution is increased, the power and cooling requirements also increase. This density needs to be balanced since there are also limits to the amount of power and cooling that are supplied to each rack. This is especially true for deep learning solutions since the massive GPU deployments usually radically increase the power and cooling requirements.

Solution overview

The solution design for a high-performance computing servers and network infrastructure is shown in Figure 12-1.

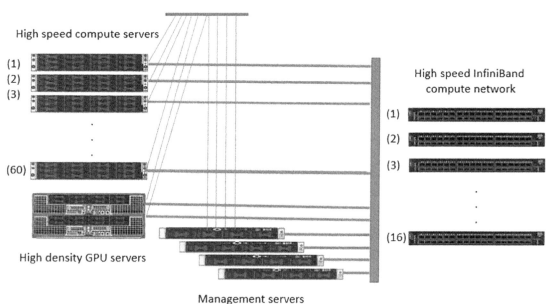

Figure 12-1 High-performance computing design

This solution uses the following components to form a "fat-tree" 56 Gb per second low-latency, high-speed compute network:

- Sixty HPE ProLiant Apollo 2200 chassis with 230 high-performance computing servers
- A single HPE Apollo 6500 chassis with two XL270d servers containing eight NVidia P100 GPUs
- Four HPE ProLiant DL360 Gen9 servers acting as management servers connected through 16 Mellanox Enhanced Data Rate (EDR) InfiniBand 36-port switches.

A network designed as a "fat-tree" logically looks like an inverted tree. The thin network pipes from the individual servers (known as leaf nodes) "tree up" into denser network pipes until it reaches the trunk (also known as the spine), which is a small number of switches that route all traffic. The Apollo chassis complex consists of 192 standard memory nodes, 28 high-memory nodes, and 10 2-GPU nodes and 2 8-GPU nodes for deep learning.

Originally designed with ProLiant Gen9 servers, a natural upgrade path for this configuration would be to Gen10 servers. ProLiant Gen10 servers would provide the following advantages to a high-performance cluster with GPUs:

- Two additional memory channels,
- 2666 MHz memory speed,
- Support for additional PCI Express 3 lanes, and
- AVX2 instruction set.

Hardware components

The following is a detailed listing of the hardware components and specifications included in the design.

- 60x Apollo 2200 chassis:
    - 192x Standard memory XL170r servers
        - Dual E5-2650v4 processors
        - 128 GB RAM
        - InfiniBand EDR port
        - 4TB SATA drive
        - 1 Gb Ethernet Management port
    - 28x High memory XL170r servers
        - Dual E5-2650v4 processors
        - 512 GB RAM
        - InfiniBand EDR port
        - 4 TB SATA drive
        - 1 Gb Ethernet Management port
    - 10x GPU XL190r servers
        - Dual E5-2650v4 processors
        - 256GB RAM

- InfiniBand EDR port
- 4 TB SATA drive
- 1 Gb Ethernet Management port
- Two NVidia K80 GPUs
- 1x Apollo 6500 chassis
    - 2x Deep learning XL270d servers
        - Dual E5-2690v4 processors
        - 512GB RAM
        - InfiniBand EDR port
        - Dual 150GB SATA SSD drives
        - 4x 1.92TB SATA SSD drives
        - 1 Gb Ethernet Management port
        - Eight NVidia P100 12GB GPUs
- 4x DL360 Gen9 management server seach with:
    - Dual E5-2650v4 processors
    - 32GB RAM
    - InfiniBand EDR port
    - 2x 1TB SATA drives
    - 1 Gb Ethernet Management port
    - 10 Gb Ethernet workload port

How did we arrive at this solution?

This section covers three key considerations in creating a design for this solution. The three topics covered in this section are server space, server power, and server cooling.

Server space

The Apollo 2200 solution was considered ideal for the following reasons:

- The Apollo 2200 had an internal capacity to carry either four XL170r dual processor servers with standard or high-memory compute capacity, or two XL190r servers with GPU capacity. This incredible capacity for compute was contained in 2U of rack space.

- Disk requirements were small, as is typical in high-performance compute environment. A single 4 TB disk was configured in every server.
- The Mellanox EDR InfiniBand card provided 56 GB per second of low-latency network capacity.
- The management network was accommodated within each server as part of the solution.

The Apollo 6500 solution was consider ideal for the following reasons:

- The Apollo 6500 contains up to 8 GPUs in a 2U space. For deep-learning projects, this provides 1792 64-bit cores per GPU, or 14336 cores per server.
- The Mellanox EDR InfiniBand card provided 56 GB per second of low latency network capacity.
- The management network was accommodated within each server as part of the solution

There are many ways to address the space requirements of this project. The following options were also considered:

- Standard rack-mount servers

    This option would consume twice the space of the Apollo 2200 solution and was quickly dismissed.

- Apollo 6000 servers

    This option provides similar density, but posed an immediate drawback. The racks provided in the data center were already populated with power distribution units that could not be used by the Apollo 6000 system.

- Apollo 8000 servers

    This option also provides excellent density and superior power utilization; however, the data center could not accommodate liquid cooling.

All things considered, the Apollo 2200 and Apollo 6500 proved to be the ideal solutions for the reasons outlined above.

Server power

A key consideration in this solution was maintaining the power envelope and power configuration that the data center had for maximum power that could be used within a rack.

After running several solutions through the HPE Power Advisor, it was determined that the racks could accommodate no more than 18 Apollo 2200 chassis. As stated above, the Apollo 6000 solution could not use the power distribution units supplied, and the Apollo 8000 could not be used since liquid cooling was not an option.

CHAPTER 12
High Performance Computing

The Apollo 6500 provided a unique challenge. This server design contains an individual power shelf. All other components of this solution required two 30Amp power sources to provide full redundancy. However, due to the high-power draw of the NVidia GPUs, this portion of the solution required three 30Amp circuits.

An example output from HPE Power Advisor for one rack of this configuration is shown in Figure 12-2.

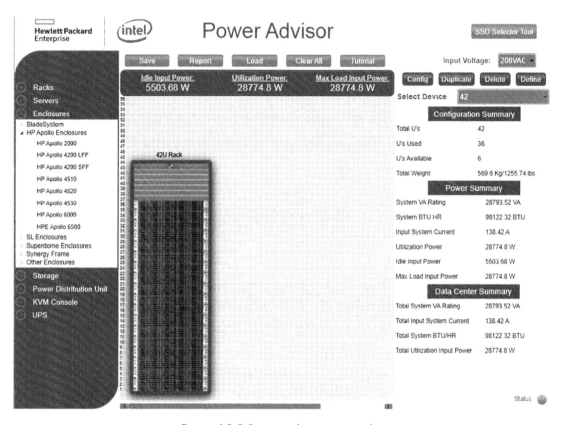

Figure 12-2 Power advisor screen shot

Server cooling

Densely designed HPC solutions can generate tremendous amounts of concentrated heat when operated at capacity. As shown in the above example output from HPE Power Advisor, that single rack portion of the cluster highlighted above generates almost 100,000 BTUs/hour! This heat must be dissipated quickly to avoid damaging any equipment within the rack or any equipment located near that rack. The final key consideration in this solution is ensuring that the environmental conditioning is in place to disseminate the heat that is generated by this cluster.

The data center used forced air cooling. Since the cluster equipment presumes front-of-rack to back-of-rack airflow, venting tiles were placed in front of racks. Air temperature was measured at floor level, and at the top of rack, to assure that cool air was available for all servers within the rack.

Hardware inventory

The following series of tables show the Bill of Materials (BOM) for the individual hardware components of this solution.

Table 12-1 BoM for a single Standard Memory (128GB) Apollo 2200

Qty	Product	Description
1	798152-B21	HP Apollo r2200 12LFF CTO Chassis
4	798155-B21	HP ProLiant XL170r Gen9 CTO Svr
4	850306-L21	HPE XL1x0r Gen9 E5-2650v4 FIO Kit
4	850306-B21	HPE XL1x0r Gen9 E5-2650v4 Kit
32	836220-B21	HPE 16GB 2Rx4 PC4-2400T-R Kit
4	825110-B21	HPE IB EDR/EN 100Gb 1P 840QSFP28 Adptr
4	846783-B21	HPE 4TB 6G SATA 7.2K LFF 512e LP MDL HDD
4	798178-B21	HP XL170r/190r LP PClex16 L Riser Kit
4	798182-B21	HP XL170r Gen9 LP P2 PClex16 R Riser Kit
4	800060-B21	HP XL170r Mini-SAS B140 Cbl Kit
2	720620-B21	HPE 1400W FS Plat Pl Ht Plg Pwr Spply Kit
2	611428-B21	HP DL2000 Hardware Rail Kit

Table 12-2 BoM for a single high-memory (512 GB) Apollo 2200

Qty	Product	Description
1	798152-B21	HP Apollo r2200 12LFF CTO Chassis
4	798155-B21	HP ProLiant XL170r Gen9 CTO Svr
4	850306-L21	HPE XL1x0r Gen9 E5-2650v4 FIO Kit
4	850306-B21	HPE XL1x0r Gen9 E5-2650v4 Kit
64	805351-B21	HPE 32GB 2Rx4 PC4-2400T-R Kit
4	825110-B21	HPE IB EDR/EN 100Gb 1P 840QSFP28 Adptr
4	846783-B21	HPE 4TB 6G SATA 7.2K LFF 512e LP MDL HDD
4	798178-B21	HP XL170r/190r LP PClex16 L Riser Kit
4	798182-B21	HP XL170r Gen9 LP P2 PClex16 R Riser Kit
4	800060-B21	HP XL170r Mini-SAS B140 Cbl Kit
2	720620-B21	HPE 1400W FS Plat Pl Ht Plg Pwr Spply Kit
2	611428-B21	HP DL2000 Hardware Rail Kit

CHAPTER 12
High Performance Computing

Table 12-3 BoM for a single GPU (256GB with two NVidia K80 GPUs) Apollo 2200

Qty	Product	Description
1	798152-B21	HP Apollo r2200 12LFF CTO Chassis
1	HA453A1-003	HP Fctry Express Blade Svr Pkg 3 SVC
2	798156-B21	HP ProLiant XL190r Gen9 CTO Svr
2	850306-L21	HPE XL1x0r Gen9 E5-2650v4 FIO Kit
2	850306-B21	HPE XL1x0r Gen9 E5-2650v4 Kit
16	836220-B21	HPE 16GB 2Rx4 PC4-2400T-R Kit
2	825110-B21	HPE IB EDR/EN 100Gb 1P 840QSFP28 Adptr
2	846783-B21	HPE 4TB 6G SATA 7.2K LFF 512e LP MDL HDD
2	798178-B21	HP XL170r/190r LP PCIex16 L Riser Kit
2	827353-B21	HP XL190r Gen9 GD-RT R Riser Kit
2	802855-B21	HP XL190r Gen9 Dul K80/M60 GPU Adptr Kit
2	800061-B21	HP XL190r Mini-SAS B140 Cbl Kit
4	J0G95A	HPE NVIDIA Tesla K80 Dual GPU Module
2	800807-B21	HP XL190r Gen9 Nvidia Enablement Kit
2	720620-B21	HPE 1400W FS Plat Pl Ht Plg Pwr Spply Kit
2	740713-B21	HP t2500 Strap Shipping Bracket

Table 12-4 BoM for a single XL270d Gen8 8-GPU server

Qty	Product	Description
1	845628-B21	HPE XL270d Gen9 Node CTO Svr
1	845628-B21 0D1	Factory integrated
1	853942-L21	HPE XL270d Gen9 E5-2690v4 FIO Kit
1	853942-B21	HPE XL270d Gen9 E5-2690v4 Kit
1	853942-B21 0D1	Factory integrated
16	805351-B21	HPE 32GB 2Rx4 PC4-2400T-R Kit
16	805351-B21 0D1	Factory integrated
1	825110-B21	HPE IB EDR/EN 100Gb 1P 840QSFP28 Adptr
1	825110-B21 0D1	Factory integrated
1	665249-B21	HPE Ethernet 10Gb 2P 560SFP+ Adptr
1	665249-B21 0D1	Factory integrated
1	869374-B21	HPE 150GB SATA 6G RI SFF SC DS SSD
1	869374-B21 0D1	Factory integrated
2	872352-B21	HPE 1.92TB SATA 6G MU SFF SC DS SSD

(Continued)

Table 12-4 BoM for a single XL270d Gen8 8-GPU server—cont'd

Qty	Product	Description
2	872352-B21 0D1	Factory integrated
1	850508-B21	HPE XL270d Gen9 4:1 Module Riser Kit
1	850508-B21 0D1	Factory integrated
1	852066-B21	HPE XL270d Gen9 Mini SAS B140 Cbl Kit
1	852066-B21 0D1	Factory integrated
8	Q0E21A	HPE NVIDIA Tesla P100 PCIE 16GB Module
8	Q0E21A 0D1	Factory integrated
2	852162-B21	HPE XL270d Gen9 NVIDIA GPU Enable Kit
2	852162-B21 0D1	Factory integrated

Table 12-5 BoM for a single DL380 Gen9 management servers

Qty	Product	Description
1	755259-B21	HP DL360 Gen9 4LFF CTO Server
1	818178-L21	HPE DL360 Gen9 E5-2650v4 FIO Kit
1	818178-B21	HPE DL360 Gen9 E5-2650v4 Kit
4	805347-B21	HPE 8GB 1Rx8 PC4-2400T-R Kit
2	652753-B21	HP 1TB 6G SAS 7.2K 3.5in SC MDL HDD
1	766211-B21	HP DL360 Gen9 LFF P440/H240 SAS Cables
1	789388-B21	HP 1U LFF Gen9 Mod Easy Install Rail Kit
1	761872-B21	HP Smart Array P440/4G FIO Controller
1	825110-B21	HPE IB EDR/EN 100Gb 1P 840QSFP28 Adptr
2	720478-B21	HPE 500W FS Plat Ht Plg Pwr Supply Kit
1	339778-B21	HP Raid 1 Drive 1 FIO Setting

Table 12-6 BoM for Mellanox switches

Qty	Product	Description
1	834976-B21	Mellanox IB EDR 36P Unmanaged Switch
1	834978-B21	Mellanox IB EDR 36P Managed Switch

The software used in this solution is call custom, and related to solving the HPC problem, so there is no software included in this BOM.

Implementation project plan

A primary goal for successfully installing a high-performance compute solution into an existing environment is accurately and precisely placing equipment for optimum space utilization, power connection and consumption, cooling benefit, and network cable length and connectivity.

The design team used a floor placement and power model of the rack distribution within the datacenter. Then the distribution of Apollo 2200 and DL360 chassis deployment within the five racks was calculated. This was adjusted to account for the following:

- Mellanox InfiniBand switch distribution
- Maximum power available within each rack
- Available cooling within each rack

At that point, an elevation document was created that shows the correct Ethernet, InfiniBand, and power cable lengths necessary to reach from the servers to the equipment. This paved the way for engineering experts to come onsite to perform installation using the rack distribution diagram and labelling guide for a fast and efficient method for the systems to be installed, cabled, labeled, and powered on.

Given the high degree of planning, this solution was installed very quickly. Within days of installation, all servers had been successfully powered on and communication within nodes and within the management infrastructure had been achieved.

Summary

As underlying components change, such as processors and memory, and application requirements increase, this solution will require modification. At the time of this writing, Gen10 components are being integrated into the environment to introduce the advantages of this platform as touched on earlier in the chapter and covered in detail in Chapter 4.

This solution is designed to accommodate growth. Based on the high degree of granularity that the Apollo 2200 allows, the growth can occur rapidly or in small increments. This high-performance compute cluster can still grow in compute power, connectivity, and management. Deep-learning technology is in its infancy, and very large changes are expected.

However, these calculations must continually be revisited and revised. Solutions of this type typically have a lifespan of two to three years at most before new technology innovations and increased application requirements outstrip the fundamental limitations of current architectures, and the entire solution set must be revisited.

13 HPE Nimble: Mixed Virtualized Workloads

INTRODUCTION

The HPE Nimble storage solution covered in this chapter enables a broad consolidation of mixed virtualized workloads.

Design considerations

While server virtualization created opportunities for consolidation and lowering costs, it also introduced some storage-related issues. Once virtual infrastructures achieved broader adoption, storage started to become a significant bottleneck due to the Input/Output (I/O) blender effect. The I/O blender effect is the mixing of multiple I/O requests or varying sizes from multiple Virtual Machines (VMs) residing on the same host over the same wire to a single target. Even workloads that have been optimized for sequential I/O now became randomized. The end result is lower performance and higher latency.

Broadly speaking, application workloads are characterized either as On-Line Transaction Processing (OLTP) or On-Line Analytical Processing (OLAP) also known as Business Intelligence (BI) workloads. OLTP is characterized by many short online transactions such as INSERT, DELETE, and UPDATE. OLTP is measured by the number of transactions per second. The following are some key OLTP factors:

Transactional workloads typically consist of

- Large number of users
- Random I/O patterns
- Small block I/O sizes
- Atomically, querying and modify small amounts of data
- High I/Os per second
- Low response time

Additionally, transactional workloads place an emphasis on data availability, data protection, and data integrity since these are typically, customer-facing, revenue-generating applications.

BI, on the other hand, is characterized by a low number of transactions that are complex and include aggregation. There is typically a lot of historical data and data mining used in BI. The following are some key BI factors:

Business Intelligence workloads consist of:

- Small number of users
- Complex and resource-intensive queries
- Fewer but larger I/O size requests
- High bandwidth requirements

These mixed environments create a storage challenge that will be addressed in the next section using an HPE Nimble all-flash solution.

Solution requirements

The solution addresses all of there quisite parameters shown in Table 13-1.

Table 13-1 Solution parameters

Capacity and data protection requirements	
Effective usable capacity	100TiB
Dataset size	65TiB
Annual growth %	20%
Data compression	Enabled
Data deduplication	Enabled
Data protection	Triple parity
Workload and I/O requirements	
Minimum IOPS	100,000
IOPS/effective usable TB	1000
Response time	<1ms
Total reads %	70%
Total writes %	30%
Sequential reads %	20%
Sequential writes %	10%
Random read size	16 KB
Random write size	8 KB
Sequential read size	64 KB
Sequential writes size	128 KB

TOOLS AND TECHNOLOGIES TO ACCELERATE DIGITAL TRANSFORMATION
Hybrid IT and Intelligent Edge Solutions

Solution overview

To achieve the stated requirements and objectives, a Nimble Storage All Flash, AF-9000 solution was configured as shown in Figures 13-1 and 13-2.

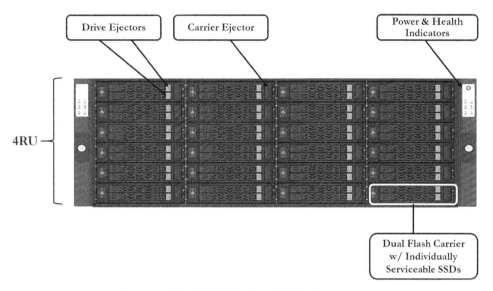

Figure 13-1 AF-9000 4U-48 SSD Front layout

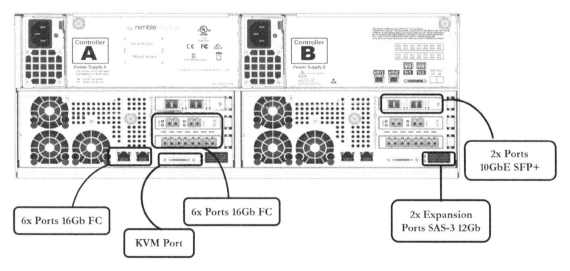

Figure 13-2 AF-9000 Rear layout

CHAPTER 13
HPE Nimble: Mixed Virtualized Workloads

This system has the following characteristics:

- AF-9000
- 4U x 48 Slots
- 12Gb SAS backplane
- Dual Intel E5-2699v3 (18 cores per socket)
- Dual 10Gb-T (Management)
- Dual 10Gb SFP+ (Replication)
- 12x 16Gbit FC Ports
- 2x 8GB NVDIMM (DDR4)
- Raw capacity: 51.84TB/47.29TiB SSD
- Effective usable capacity: 112.14TB/101.99TiB
- All inclusive SW
    o Triple parity and intra-drive parity (Triple+ RAID)
    o Snapshots – 300,000 supported maximum
    o Variable block inline deduplication
    o Variable block inline compression
    o Inline zero block removal
    o Replication
    o Encryption—AES-256bit FIPS 140-2 validated
    o Oracle Snap & Clone Integration
    o Microsoft Aware Snap & Clone Integration
    o VMware Integration
    o Nimble connection manager for VMware

- o Automated quality of service (QoS)
- o User-defined quality of service (QoS) Controls
- o Full representational state transfer (REST) API
- o Artificial intelligence (AI) and predictive analytics using HPE-Nimble InfoSight
- o Six Nines Data Availability Guarantee

Scaling capabilities

Given the dynamic nature and frequent bursts in data of this virtualized environment, the platform's scale-to-fit design allows us to start with a modest configuration and then increase the capacity when required. The system can scale-up its capacity and performance independently in an online, nondisruptive manner.

The platform can also scale performance and capacity beyond a single array into a scale-out group.

The scale-out group can be comprised of multiple pools using the same or even different model arrays with nondisruptive data mobility options and is managed as a single unit.

Alternatively, the group can be deployed as a single striped pool that aggregates compute and memory resources and is managed as a single performance and capacity entity.

Within each physical host, it is recommended that the Nimble Connection Manager (NCM) is installed for either Fibre Channel or iSCSI deployments. NCM manages the appropriate number sessions from host to the storage group, sets the recommended Fibre channel driver timeouts, and sets there commended Path Selection Policy (PSP) for the platform.

Management capabilities

The storage system comes with an embedded, easy-to-use, array management tool. This uses an intuitive HTML5 graphical user interface (GUI) from which to manage the array as shown in Figure 13-3.

CHAPTER 13
HPE Nimble: Mixed Virtualized Workloads

Figure 13-3 Example of Array Management GUI

The GUI can be accessed via any supported browser (IE, Firefox, and Safari) using login and password credentials established during the array setup. The six main areas listed on the dashboard are Manage, Hardware, Monitor, Events, Administration, and Help. Under each area, there are subareas providing granular monitoring and management controls. Use the menu items to move to any location of interest.

Advanced analytics

HPE InfoSight (referred to here as InfoSight) is a cloud-based SaaS portal available as part of the core architecture, designed to dramatically improve not only storage system reliability but also extend this reliability across the entire infrastructure stack. InfoSight can monitor, alert, and provide actionable recommendations based on end-to-end correlational analysis of the entire install base. Through powerful data science, InfoSight completely transforms a typically reactive, error-prone support, and management experience into a proactive process for maintaining peak storage health. The following section highlights those advanced analytics capabilities in greater detail.

InfoSight predictive analytics

Infrastructure complexity, configuration variability across applications, hosts, hypervisors, and virtual machines have made infrastructure disruption all but inevitable. To combat this issue, Nimble Storage took a unique approach and began embedding diagnostic sensors into every module of code, building a foundation for deep health, performance analytics, and workload characterization along with an AI recommendation engine.

Figure 13-4 and 13-5 show how InfoSight-based analytics can be leveraged to gain a meaningful understanding of application workloads by identifying I/O patterns, recurring performance patterns, and noisy neighbor workloads.

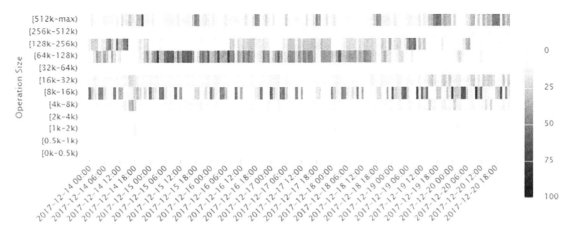

Figure 13-4 InfoSight—Oracle Operation Size Heatmap

Figure 13-4 shows the distribution of all I/O operations (reads and writes) across a range of block sizes. The darker the region, the larger the proportion of the total operations that fell into that size range at that time. For each range of I/O sizes, the square bracket denotes a closed interval endpoint and the parenthesis denotes an open interval endpoint (for example, [8k–16k) is inclusive of 8k operations but does not include 16k operations). In this particular example, we see that most I/O operations occur at the ranges of [8k-16k) and [64k-128k).

CHAPTER 13
HPE Nimble: Mixed Virtualized Workloads

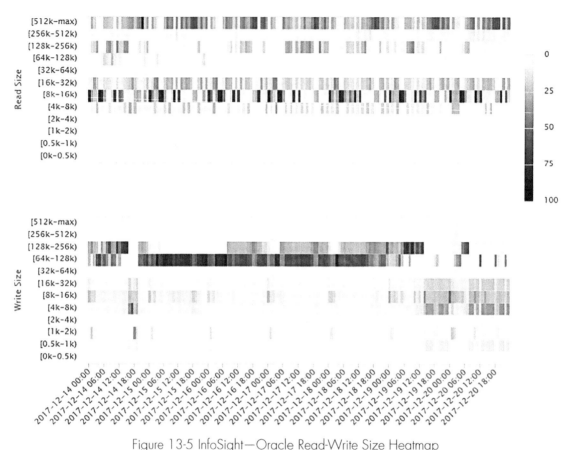

Figure 13-5 InfoSight—Oracle Read-Write Size Heatmap

Figure 13-5 separates application I/O distribution into read and write across a range of I/O sizes. The darker the region, the larger the proportion of read or write operations fell into that size range at that time. For example, although we see that Oracle reads are distributed across a wide range of I/O sizes, we notice that the largest proportion of reads falls into a [8k-16) range. Similarly, we notice that the largest proportion of writes falls into [64k–128k) range.

Recurring performance patterns identify application intensity at five-minute intervals for a period of one month. As shown in Figure 13-6, Oracle is very active between 18:00 and 19:15 every day indicating possible end-of-day processing or backup activity. Using recurring performance patterns, we can also identify low I/O intensity periods and high-intensity jobs can be scheduled.

Hybrid IT and Intelligent Edge Solutions

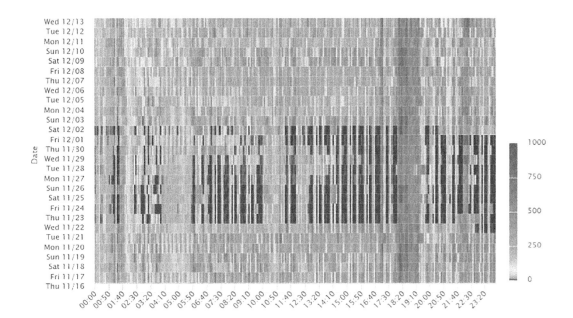

Figure 13-6 InfoSight—Oracle Recurring Performance Patterns

Intervolume performance and contention shows volumes consuming most I/O resources and volumes that are impacted by high latency as a result of volumes that are consuming most resources. This is the proverbial "noisy neighbor" scenario.

As we see in Figure 13-7 volume "*colo-asup-tpsanstats1*" and "*colo-asup-san1*" are experiencing high latency due to high I/O activity by volume "*colo-asup-sanstats1.*" If the affected volumes have reached unacceptable latency levels, then volume "*colo-asup-sanstats1*" is a Quality of Service (QoS) candidate for IOP and/or throughput limits.

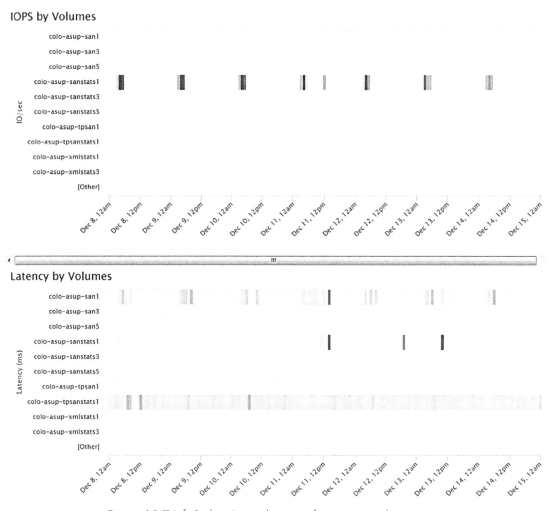

Figure 13-7 InfoSight—Intervolume performance and contention

Hardware inventory

Table 13-2 shows the Bill of Materials (BOM) for the Nimble Storage AF9000 storage array. As with the processors and NAND Flash densities, technology has advanced since the time of this writing so there are newer models available.

Table 13-2 Solution BOM

Qty	Product #	Product description
1	Q8B35A	HPE Nimble Storage AF9000 All Flash Dual Controller 10GBASE-T 2-port Base Array
1	Q8B70A	HPE Nimble Storage AF All Flash Array 24x240GB Flash Bundle
1	Q8B70A 0D1	Factory integrated
1	Q8C08A	HPE Nimble Storage 2x10GbE 2-port and 2x16Gb FC 4-port and 2x16Gb FC 2-port Adapter Kit
1	Q8C08A 0D1	Factory integrated
2	Q8F97A	HPE Nimble Storage C13 to C14 250V 13Amp 1m PDU Base Array Power Cord
2	Q8F97A 0D1	Factory integrated
1	Q8G27A	HPE Nimble Storage NOS Default Software
1	Q8G27A 0D1	Factory integrated
1	Q8G61A	HPE Nimble Storage AF5000/7000/9000 All Flash Array 24x1.92TB Flash Bundle
1	Q8G61A 0D1	Factory integrated
1	HA114A1	HPE Installation and Startup SVC
1	HA114A1 5MR	HPE NS Array Startup SVC
1	HT6Z2A3	HPE NS 3Y FC 4H Onsite Exchange Support
1	HT6Z2A3 X4N	HPE NS AF9000 All Flash Base Array Supp
1	HT6Z2A3 W6C	HPE NS AF5/7/9000 All Fsh 46TB Fsh Supp
1	HT6Z2A3 X43	HPE NS AF All Flash 5.76TB Flash Supp
1	HT6Z2A3 X65	HPE NS 10GbE 2p/16Gb FC 4p/2p Adp Supp

Summary

The Nimble storage solution covered in this chapter is built to deliver the data capacity, data protection, and I/O requirements to enable a broad consolidation of mixed virtualized workloads. The solution features purpose-built all-flash arrays with always-on inline data reductions that provide a consistent application performance experience, even under adverse conditions. The system can scale-up capacity and performance independently in a nondisruptive manner and has powerful management capabilities complete with HPE InfoSight's predictive analytics designed to monitor and maintain peak storage health. QoS controls can be both internally automated as well as user-defined. The end result is a system with lower operational risk, higher performance, and increased capacity that can be seamlessly scaled and easily managed to protect your storage investment now and for the future.

14 High-Performance SAN and All-Flash

INTRODUCTION

This chapter covers a variety of topics related to high-performance Storage Area Network (SAN) including background and example designed for small, medium, and large SANs.

Many workloads require a high-performance Storage Area Network (SAN) implemented with Fibre Channel (FC) for a variety of reasons including, but not limited to, the following list:

1. Availability—reliability and resiliency
2. Scalability—support large-scale deployments
3. Performance—speed, throughput, IOPS, and latency
4. Extensibility—adaptability to future technology enhancements
5. Manageability—simplified management based on integrated tools and automation
6. Security—prevention of unauthorized access, modification, and misuse of network resources
7. Agility—ease of scale to accommodate growth
8. Acquisition and Ownership costs—initial and ongoing costs to operate

 Note

There are a number of additional storage solutions not covered in this chapter that solve a variety of different problems other than that of a high-performance SAN. These include, for example, Object Storage for large files and Network Attached Storage (NAS). The SimpliVity chapter covers storage for Virtual Machines (VMs) and containers that is both compressed and de-duplicated which is ideal for hosting many VMs with easy management.

Fibre Channel is unique in that it is the best and only purpose-built storage networking choice when it comes to shared storage, server virtualization, and flash storage. In today's environment, Gen 6 (32/128 GFC) is the minimum requirement for all flash arrays and high-performance computing. The upcoming "SAN" section has a bullet list covering the advantages of Gen 6. As you start to evaluate implementing a FC SAN, consider Gen 6 as the starting point. In this chapter, we are going to focus on the FC SAN and low-latency high-performance (shared) storage arrays using Solid State

Drives (SSD,) FC-Non Volatile Memory Express (NVMe), and hybrid (SSD/HDD) storage, namely, All-Flash Array (AFA) or Hybrid-Flash Arrays (HFA).

SAN considerations and overview

There are a variety of technical components of which this solution is comprised that are covered in the upcoming sections.

Storage

All-Flash Arrays (AFA) are the best option to achieve the highest performance available. You have a variety of choices for all flash that are covered in this section, including:

- HPE 3PAR StoreServ built to meet the extreme requirements of massively consolidated cloud service providers.
- HPE Nimble Multicloud flash fabric, an easy-to-use, single flash architecture that can be used across a multi-cloud environment.

In the midrange to high-end, the **HPE 3PAR StoreServ** family of flash-optimized data storage systems modernizes your data center to instantly handle unpredictable workloads with 99.9999% data availability and up to 1.2 million Input Output Operations/second (IOPs) performance. Features include rapid and automated provisioning, multi-tenant design, hardware-accelerated deduplication and compression, and sub-1ms latency, all within a single tier-1 storage architecture that starts small and scales big.

Nimble Storage is radically simple, cloud-ready, all-flash storage solution that leverages the power of predictive analytics. The Nimble Storage AF1000 All Flash Array is the entry point to the Nimble Storage all-flash product line.

A third option is the hybrid flash **HPE MSA** 2052 Storage system designed for affordable application acceleration ideal for small and remote office deployments. The system starts with 1.6 TB of flash capacity and can be scaled as needed with any combination of enterprise-level and high-performance SSDs. Delivering performance in excess of 200,000 IOPS, the MSA 2052 Storage system can save you up to 40% with an all-inclusive software suite and 1.6 TB of flash capacity included.

Storage Area Network (SAN)

A SAN will maximize the performance and ROI of an all-flash data center with a Gen 6 FC network. A FC SAN will maximize ROI and protect the customers' investments because it does not require a redesign of the ecosystem with any special hardware or management upgrades. Fibre Channel is the preferred infrastructure choice for high-performance AFA storage, hybrid, or even HDD arrays. According to industry analysis, greater than 80% of SSD storage is deployed on Fibre Channel SANs. This is primarily because of FC's low-latency and high IOPs performance, security, data integrity, and stability. As more deployments of flash storage and Non-Volatile Memory Express (NVMe) flash storage become prevalent, industry professionals are recommending Gen 6 FC (32/128 GFC) as the minimum network requirement. The upcoming "SAN" section lists the main advantages of Gen 6.

Brocade portfolio spans from 8 port entry rack switches for small companies to very high port (>300) Directors used for large data centers. Depending on your use case, you can choose any combination, from rack mount (or edge switch) for top-of-rack or smaller deployments, SAN extension for disaster recovery, embedded FC or Virtual Connect switch modules for Synergy Frame or BladeSystem server enclosures, and SAN Analytics to monitor application end-to-end performance and predictable performance. The document "SAN Design Reference Guide" provides a lot of detail on this process and can be accessed here: https://support.hpe.com/hpsc/doc/public/display?docId=c00403562

The following list describes some of the features of Gen 6 and indicates how it can be considered an appropriate starting point for implementing FC SAN:

- Always on architecture: proven, redundant air-gapped fabrics
- Flash ready performance: low-latency, high bandwidth connectivity
- Simple scalability: petabyte scale for tens of thousands of servers and storage
- Integrated security: secure by design with isolated network and managed access
- Typical workloads: mission critical databases, hypervisors, VDI, OLTP, ERP, and CRM

Figure 14-1 shows a basic SAN configuration.

CHAPTER 14
High-Performance SAN and All-Flash

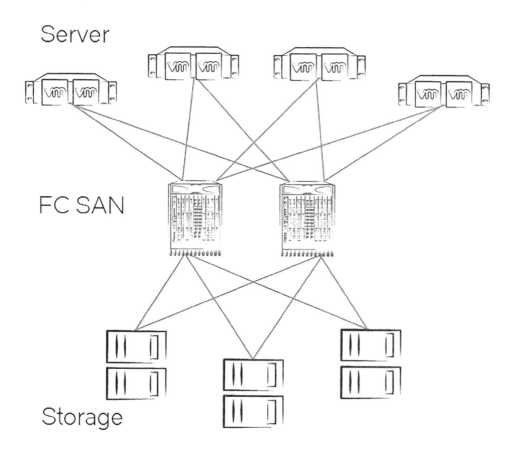

Figure 14-1 Basic Two-Director FC SAN configuration

The next section covers NVMe over FC considerations.

NVMe over Fiber Channel (NVMe-oFC)

An NVMe SSD all-flash array on an FC network will help transform IT and deliver more value back to the business, latency improvements of up to 55% over traditional SAN flash storage will unleash the full capabilities of NVMe storage.

This scalable storage network technology will deliver reliable flash that performs at the speed of memory over FC Fabrics.

A positive for NVMe over FC is that it supports multiprotocol over FC, that is, SCSI, FICON, and FC storage. Thus, there is no need to worry about having to rip-replace your current shared storage or add a separate FC SAN. You can run all three over the same fabric.

Here are some further considerations favoring NVMe over FC:

- Application transactions will accelerate because AFA are 100x faster than arrays with spinning HDDs, which will provide a better user experience.
- Databases will be able to increase the number of queries they support, meaning lower latency + lower overhead = higher IOPS
- More types of workloads can be consolidated on hypervisors due to the improved storage performance
- Plug-and-play operation with any existing Gen 6 FC 16G/32G storage network and HPE StoreFabric SN1200E/SN1600E host bus adapters. There is no rip-and-replace required. Just connect an NVMe storage and go.

Gen 6 FC is gold standard and future proofed, ready today for when for NVMe storage becomes readily available.

Compute in this solution is covered in the next section.

Compute

We describe High Performance Compute (HPC) Servers in the High Performance chapter along with the other server technology available to meet your application needs. These choices are applicable and important to understand the relationship between compute, network, and storage bandwidth considerations. However, there is no right or wrong server compute choice. All things being equal, the make or model of the compute platform is not relevant when implementing a FC SAN. What is essential is that the server's compute resources meet your customers' business needs.

Server Host Bus Adapter (HBA)

The HBA is the "on ramp" to an FC SAN. Evolving with each new generation of FC, the latest Gen 6 HPE StoreFabric HBAs, such as the Emulex SN1200E and SN1600E models, support 16G and 32G FC, respectively.

The key to better application outcomes is ensuring that the HBA can deliver the necessary throughput, transaction rates, and response time for optimum application performance. You should also consider having enough surplus bandwidth overhead to ensure your fabric can grow without interruption. If you are not sure what 'enough' really means, take a closer look at your current fabric workload and extrapolate a plan based on two three-year growth.

The SN1200E and SN1600E both deliver up to 1.6 million IOPs, thanks to the Emulex HBA's Dynamic Multi-core ASIC architecture. This ensures that the server I/O is no longer a constraint for maximum application performance. In addition to sheer performance, other reasons that make the SN1200E/SN1600E the right choice for the FC SAN include:

CHAPTER 14
High-Performance SAN and All-Flash

- **Investment protection.** The SN1200E/SN1600E are the only HPE StoreFabric FC HBAs that are "NVMe Enabled" today that will only require driver and firmware upgrade to make them compliant with the industry-ratified, NVMe over Fibre Channel specification.

- **Secure.** The SN1200E/SN1600E protect against "rogue firmware" with an Emulex exclusive feature that verifies the digital signature of the firmware before downloading.

- **Simplified and accelerated** SAN deployment with Emulex HBAs that are compatible with HPE Smart SAN for 3PAR. Please see the Management section below for additional information.

Applications

Applications drive the need for HPC, FC SAN, and all-flash storage. With the introduction of NVMe Fabrics and all-flash arrays, these new technologies are driving the modernization of storage architectures. At the top of the architecture are applications driving the data center requirements.

What is very interesting to note in the next illustration (Figure 14-2) is that Emulex-Broadcom have kept the same servers and only changed out the HBA's to see a significant I/O impact across different application workloads. They have proved that simply by changing out the generation of HBA from Gen 5 8G FC to Gen 6 16G FC. The application workload completion time was reduced by 2x. Changing the generation of the HBA from Gen 5 8G FC to Gen 6 32G FC showed a 4x improvement.

Figure 14-2 TPC-H Benchmarks from Emulex-Broadcom

SAN extension

Most Disaster Recovery (DR) plans involve multiple data centers for failover purposes. A more common solution is two data centers operating in active-active mode. Active-Active is where two data centers are designated failover sites for each other. When deploying a FC SAN, there are alternative hardware deployment options. One option is an enterprise-class FC SAN extension blade for your director class switch, or a top-of-rack Brocade 7840 that extends a SAN by using the WAN technology to bridge datacenter together. In both deployments, a dedicated architecture is better able to deliver predictable, reliable, and scalable network performance on-premises, across a metro area, or across regions (Figure 14-3).

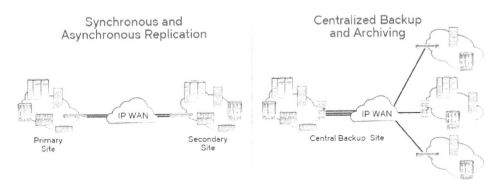

Figure 14-3 One-to-one versus many-to-one deployment

Figure 14-4 shows how the Brocade 7840 addresses the most demanding DR requirements. Twenty-four 16 Gbps Fibre Channel/FICON® ports, sixteen 1/10 Gigabit Ethernet (GbE) ports, and two 40 GbE ports provide the bandwidth, can scale up to 80 Gbps application throughput, depending on the type of data and the characteristics of the WAN connection.

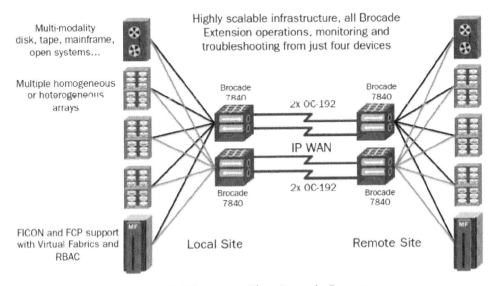

Figure 14-4 Enterprise Class Brocade Extension

Management Tools

Storage Area Networks and the tools that are used to manage them have become increasingly complex. SAN administrators spend more time in configuration tasks that are tedious, manual activities, which can potentially result in unplanned downtime. The reasons for the complexity include the increased adoption of virtualization and containerization, which is a complexity stemming from mapping the virtual and physical layers of the SAN and the adoption of multiple protocols such as iSCSI, FC, and FCoE. This section highlights some tools that can be used to simplify these tasks.

HPE SmartSAN for 3PAR StorServ

HPE Smart SAN for 3PAR makes SAN configuration and management simple and error-free through intelligent automation. It is a protocol-agnostic application embedded in SAN components that enables the 3PAR to orchestrate configuration, settings, and policies in an HPE StoreFabric SAN. HPE Smart SAN's Target Driven Peer Zoning enables you to configure zones accurately in minutes and not in hours and its automatic discovery mechanism creates a powerful platform that would enable real-time diagnostics and SAN analysis for more resiliency. HPE Smart SAN for 3PAR helps automate the labor-intensive task of SAN Zoning with the following benefits:

- Requires 80% fewer steps
- Accelerates deployments by 99%

Brocade EZSwitchSetup

For installers, Brocade has a software utility that takes the burden out of installing a switch. This installation software is complementary and can be accessed from www.mybrocade.com web site, search EZSwitch, and match your switch model to Installation guide.

Brocade Fabric Vision

Brocade Fabric Vision technology combines capabilities of the Gen 6 Fibre Channel ASIC, Brocade Fabric OS, and Brocade Network Advisor to maximize uptime, simplify SAN management, and provide unprecedented visibility and insight across the storage network. This product has diagnostic, proactive monitoring, and management capabilities, Brocade Fabric Vision technology helps administrators avoid problems, maximize application performance, and reduce operational costs.

Small, Medium, and Large SAN

This solution was selected in order to represent a range of options from a small entry point to an enterprise very high-end data center. Fibre Channel is the preferred networking protocol of choice when it comes to security, reliability, and performance. One of the primary values of a SAN is shared storage. In comparison, NAS or DAS storage is on shared Ethernet network with a one-to-one relationship with the server, whereas Fibre Channel provides many connections between storage arrays for a many-to-many relationship.

There are different entry, medium, and large high-end SAN configurations. At a minimum, we always recommend two of every component for high-availability and maximum fault tolerance. As an example, the illustration below depicts a very simplistic entry SAN comprised of one to two servers with FC host bus adapters, one to two FC switches, and at least one storage array (Figure 14-5).

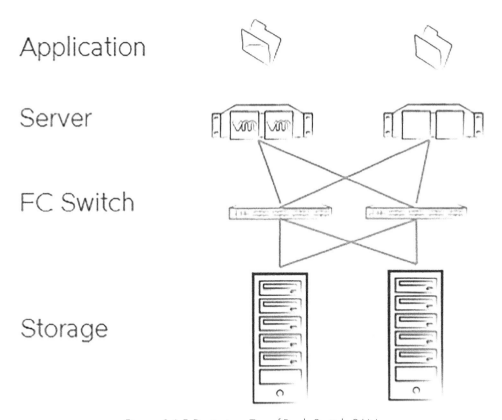

Figure 14-5 Basic two Top-of-Rack Switch SAN

A medium-size SAN is just a multiplier, adds more physical or virtual servers, and more switch ports to ensure sufficient connectivity (plus additional ports for growth) between your server platforms and storage arrays.

Very large SAN deployments can become large very quick. It may include a combination of edge Top-of-Rack switches, embedded switches in Synergy or BladeSystem servers or end-of-row aggregator switches, plus a single or multiple director class (500+ FC ports per unit) product. Add in tens to hundreds of server (physical or virtual), and petabytes-to-zettabytes of block SSD or HDD storage. The largest SAN deployments range from tens of thousands to hundreds of thousands of SAN ports for 500 to tens of thousands end-users.

There is commonality between all three of these installations, you get the same level of security, reliability, scalability, and performance. With Brocade Fibre Channel technology-based backbones and switches, you have got the firepower to deliver high-performance connectivity across the campus or across the globe. Scale your network on demand—move more data more places—as you keep costs of ownership reined in.

Small, medium, and large SAN inventory

Table 14-1 covers a small, medium, and large SAN inventory example. The small example uses a Nimble solution. See the Nimble chapter for further details. The medium implementation uses on 3PAR unit and the large example uses two 3PAR units. Part numbers change often and at the time of this writing the numbers below are current:

Table 14-1 Small, medium, and large SAN

Qty	Product #	Model	Description
Small SAN (16–48 ports)			
1	Q8B40A	AF1000	HPE Nimble Storage All Flash Array
2	Q1H70A	SN3600B	HPE SN3600B 32 Gb 24/8 FC Switch
16	P9H32A	Optics	32 Gb SFP+ SWL Optics
1	Q1Z10AAE	LTU	8-port POD LTU upgrade (optional)
2	Q0L12A	SN1600E Dual Port	HPE StoreFabric SN1600E 32 Gb Dual Port FC HBA1
	TC352B	Mgmt	HPE B-series Network Advisor Enterprise Software
Medium SAN (48–96 ports)			
Qty	Product #	Model	Description
1-2	various	7450c	HPE 3PAR StoreServ 7450 Storage
2	Q0U61A	SN6600B	HPE SN6600B 32Gb 48/48 PP+ 48p SFP+ Switch
1	TA5528AAE	License	Integrated Routing
96	P9H32A	Optics	32Gb SFP+ SWL Optics
1	TC352B	Mgmt	HPE B-series Network Advisor Enterprise Software
2	Q0L12A	SN1600E Dual Port	HPE StoreFabric SN1600E 32 Gb Dual Port FC HBA

(Continued)

Table 14-1 Small, medium, and large SAN—cont'd

Large (>96+ ports)			
Qty	Product #	Model	Description
1-2	various	7450c	HPE 3PAR StoreServ 7450 Storage
2	Q0U61A	SN6600B	HPE SN6600B 32 Gb 48/48 PP+ 48p SFP+ Switch (ToR edge, optics included)
1	Q0U63A	SN8600B	HPE SN8600B 8-slot Pwr Pack+ Dir Switch (Core)
2	Q0U86A	port blade	HPE SN8600B 32 Gb 48p SFP+ Integrated Blade (optics included)
1	Q0U85A	DR blade	HPE SN8600B 32 Gb SAN Extension Blade (optional for DR site)
1	TA5528AAE	License	Integrated Routing (for SN6600B switches)
1	Q0T81AAE	License	ICL POD LTU (required if using more than one director)
1	TC354B	Mgmt	HPE B-series Network Advisor Professional Plus Software
2	Q0L12A	SN1600E Dual Port	HPE StoreFabric SN1600E 32Gb Dual Port FC HBAServices

HPE SAN Deployment Services

Deployment services can be important in the design and implementation of a successful SAN environment. There is an HPE SAN Deployment Service that provides installation and configuration of your storage area network (SAN).

HPE SAN Deployment Service covers a comprehensive complement of technologies, including Fibre Channel (FC), Fibre Channel over Ethernet (FCoE), Fibre Channel over IP (FCIP), FICON (Fibre Channel for HPE XP Storage Array-based mainframe storage), and iSCSI or serial-attached SCSI (SAS), for switches and associated devices. This service implements a new single-fabric or dual-fabric SAN or extends an existing SAN but does not include configuration of arrays or storage devices that are covered by their own corresponding deployment services.

Summary

High-performance SAN is for the most demanding workloads that require high performance in terms of low-latency and high IOPs. This chapter covered many considerations related to high-performance components and provides example of three designs that can be tailored to meet your business needs. As with all of the topics covered in this book, feel free to contact the author at his email address in the Introduction and we can get your design underway.

15 Consumption-Based IT Services

INTRODUCTION

We live in a consumption-driven world, whether it be personally with music, television, cell phones, and travel, or professionally with real-time analytics, content storage, and data. In order to be more competitive and deliver better, faster, more reliable outcomes in a more economical manner, IT organizations need to find new and different ways to consume Information Technology (IT.) This is no longer an optional endeavor because line-of-business leaders are demanding that IT organizations advance to meet ever-changing needs. IT organizations are, therefore, looking for rapid access to agile and scalable resources, delivered with full control and a transparent and understandable utility model that allows them to pay only for what is used. These needs have resulted in consumption-based IT services. This chapter describes the underlying requirements for such services and highlights how HPE has introduced a whole suite of innovative pay-as-you-use IT solutions designed to deliver the business outcomes that top-performing IT organizations now need to drive.

What is consumption-based IT?

At its most basic level, consumption-based IT delivers IT infrastructure resources on your premises in a fluid, unrestricted, and optimal way that simplifies IT operations. It incorporates activities such as capacity management metering, resource planning, patch management, and hardware and software support. This is all delivered as-a-service allowing for ease of acquisition and operations on a pay-per-use basis. Instead of building your IT infrastructure from the ground up, consumption-based IT allows you to increase control, accelerate time to value, improve economics, reduce risk, and simplify operations through an IT-as-a-Service model, to accomplish the following:

- Start new, innovative projects without incurring upfront infrastructure costs
- Eliminate overprovisioning characteristic of most new projects and infrastructure acquisition
- Accelerate time-to-value by enabling elastic IT to deploy and scale infrastructure
- Eliminate lengthy purchasing cycles that impact your ability to quickly respond to market trends and competitive initiatives
- Support business justifications by significantly improving Return on Investment (ROI) and Net Present Value (NPV)

- Minimize the threat of "shadow IT" by supplying on-premises infrastructure with public cloud capabilities
- Increase transparency of usage and costs to improve use and profitability management
- Maintain control so that you can ensure security and compliance with on-premises deployment

What are its benefits and advantages?

Figure 15-1 shows the high-level requisite benefits and features of consumption-based IT services.

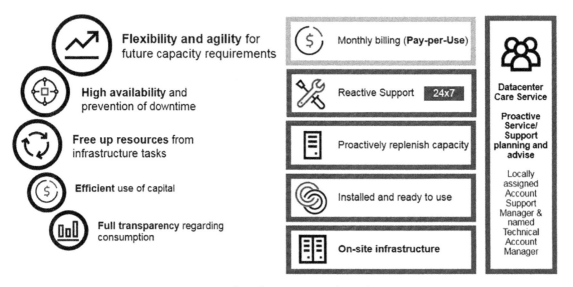

Figure 15-1 Benefits of consumption-based IT services

Staying ahead of the competition and taking advantage of new opportunities means transforming IT operations by investing in technologies that offer usage choices, flexible consumption, rapid innovation, and specific user outcomes, along with full control over your applications and data. This trend is summarized in the following quote from an IDC study discussing worldwide datacenter predictions for 2018:

> "By 2020, consumption-based procurement in data centers will have eclipsed traditional procurement through improved 'as a service' models, accounting for as much as 40% of enterprises' IT infrastructure spending."[1]

---

[1] IDC FutureScape: Worldwide Datacenter 2018 Predictions.

HPE GreenLake: Pay-per-use IT solutions

In response to these rising trends, HPE GreenLake is a flexible capacity-based suite of solutions designed to meet the growing need for on-premises, consumption-based IT. It simplifies the acquisition, deployment, and maintenance of a hybrid IT infrastructure—legacy functions, core IT, private cloud, and agile cloud-native IT—into a new, agile, usage-based model not available until now. GreenLake's Consumption and Flexible Capacity offerings are delivered by the HPE Pointnext services organization. HPE does the IT heavy lifting, and you get a simple, pay-per-use environment, maintaining full control of your apps and data in your own environment to enable:

- Rapid onboarding
- No capital outlay
- Custom designs and fully integrated architectures
- Elastic IT that evolves with your needs

Table 15-1 highlights the essential features of the HPE GreenLake consumption-based model.

Table 15-1 HPE GreenLake Essential Features

HPE GreenLake is:	HPE GreenLake is not:
✓ A **flexible consumption-based model** that speeds your time to market and grows organically as your needs change—helping you to avoid overprovisioning.	✗ A fixed IT purchase that comes with a significant capital outlay and long procurement, design, and build cycles.
✓ A **pay-as-you-go solution** that establishes a simple, outcome-oriented metric.	✗ A fixed lease with predetermined payment plans.
✓ **Operated for you**, but you remain in control. You experience no disruption, no staff transfer.	✗ An outsourced service that requires you to make an all-or-nothing decision and relinquish control.
✓ A service that takes the burden off IT operations with a **simple approach**.	✗ An appliance-like solution that is still labor intensive for IT.
✓ A **solution that delivers the advantages of public cloud services**—simplicity, scalability, pay per use—but with the security and control of on-premises IT. It also delivers more robust services that are specific to your needs.	✗ A public cloud that doesn't deliver the control and security of on-premises IT, where services are fixed or generic, and where applications must be refactored on a cloud-native architecture.

CHAPTER 15
Consumption-Based IT Services

Consumption-as-a-Service

An easy way to understand consumption-as-a-Service and Flex Capacity is to understand the design behind it. What follows is an example of the design process that includes hardware, software, and services that ultimately ends up as a consumption-based service to the customer. Figure 15-2 depicts some of the advantages of a consumption-based model:

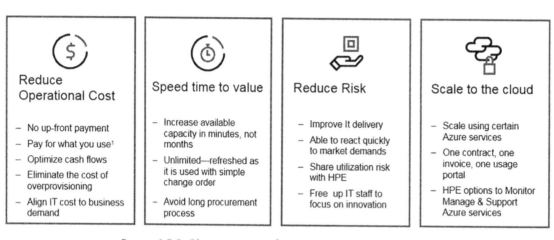

Figure 15-2 Characteristics of a consumption-based model

Flexible capacity

Many IT leaders are intrigued by the flexibility and economics of the public cloud. They enjoy the ability to scale on-demand but are frustrated with the high and rapidly escalating cost at scale and lack of control and security associated with moving environments out of their own data center. The following is a list of challenges faced by IT leaders:

- Security and control of an on-premises solution
- Reduced total ownership cost and risk
- Upfront costs and overprovisioning
- Months-long scaling on demand
- Flexibility and responsiveness to the business
- Pay for what is used as used
- Need costs to align to how the business is doing, and with visibility into what is being spent

Figure 15-3 shows the advantages of Flexible Capacity when compared to standard Capex and Opex acquisition models.

Feature	CAPEX	OPEX	Flexible Capacity
Model	Own, Build IT Ahead of Use	Build IT Ahead of Use	Use IT, Deploy As Used
Capacity Adds	Procurement/Ahead of Use	Procurement/Ahead of Use	Immediate/As Used
Infrastructure Utilization	40%-75%	40%-75%	Near-100%
Excess Infrastructure Risk	Office Depot	Office Depot	HPE
Monthly Cost	Fixed Depreciation	Fixed Lease Payment	Variable (Avg. of Daily Use)
Cost Matched to Business	Periodic Stair-Step Up	Periodic Stair-Step Up	Immediate/Up-down
Monthly Unit Cost	Variable/Unknown	Variable/Unknown	Fixed/Known
Cash Outlay	Upfront Lump Sum	Contractual Monthly	Variable Monthly Service
Tech Adaptation to Use	As-deployed	As-deployed	As Used
Tech Refresh	Manual	Manual	Automatic
Meter Report	If Acquired	If Acquired	Integral

Figure 15-3 Comparison of acquisition models

HPE GreenLake portfolio

The HPE GreenLake portfolio includes 22 packages that move from custom to standard offers in a consumption/as-a-Service market. Each package features predesigned and tested solutions including needed services to get your organization into the as-a-service market quicker. Payment is simple and based on a single pay-per-usage metric that is relevant to the particular solution and your business. The portfolio offers flexible technology options that include:

- Backup
- Infrastructure
- Big Data
- Containers
- Databases
- Microsoft Azure Stack
- SAP HANA
- Storage
- Edge
- HPC
- Compute
- *Future based solutions are already on the roadmap*

CHAPTER 15
Consumption-Based IT Services

 Note
Links to complete turnkey integrated consumption solutions can be found at the end of this chapter.

There are currently two types of Standard Packages available:

- **HPE GreenLake Flex Capacity Packages** focus on specific technologies such as HPE Synergy, HPE 3PAR, HPE ProLiant Blades. These offer typical Flex Capacity attributes such as a 3-year term, 80% commit, a buffer, and pay only for capacity used each month, customer-managed.

- **HPE GreenLake Standard Consumption Packages** deliver optimum results for specific business uses and outcomes, including Microsoft Exchange, SAP HANA, and VDI.

The HPE GreenLake portfolio includes GreenLake Consumption, where HPE will manage HPE-owned infrastructure and Flexible Capacity, where the customer manages the HPE-owned infrastructure, both of which are on-premises and designed to deliver the best outcomes with maximum efficiency and cost-effectiveness, delivering the following important advantages:

Figure 15-4 Consumption-based advantages

These advantages address some central concerns of line of business leaders, IT organizations, and executives captured in the following questions and statements:

- ***Cost optimization***

    "How can I get the most from my budgets and how can I align cost of usage to the business?"

- *Reduce risk*

    "IT is central to the business, I need to avoid downtime and ensure there is enough capacity to support business demand without holding unutilized assets."

- *Business agility*

    "We need an infrastructure that can grow with the business and enable quicker time to market."

An effective response to these needs leads to a future-state vision that embraces and extends the flexible characteristics of consumption-based IT services, including:

- Pay only as you go and grow IT business model, without investment and/or true-ups
- Rapid provisioning using simple change orders, not lengthy procurement cycles
- Ability to show-back or charge-back to Lines of Business (LoB)
- Spin-up resources in a matter of minutes, not days, or weeks
- Reduction in the number of applications
- Increase speed and integration of on premises solution(s) to public cloud
- Free-up resources for innovation
- Eliminate overprovisioning and overpaying

Sample design and implementation

This section shows a sample HPE GreenLake architecture crafted to meet these IT and business objectives, namely, an on-prem solution with the following elements and attributes:

- Self-service portal, also called a "marketplace" that allows users to deploy operating systems such as Windows and Linux
- Ability to limit options around processor, disk, and memory
- Optional resiliency including off-site replication for Disaster Recovery (DR)
- Backup and recovery capability built-in
- Integration into existing toolsets where appropriate
- Capability to deploy virtual, bare metal, and container solutions
- Metering capability to measure usage and show-back or charge-back to the business
- Reporting capability
- Financial consumption model for hardware
- Custom Docker support

CHAPTER 15
Consumption-Based IT Services

Figure 15-5 shows how all of the components can be integrated into a single unified design that leverages the flexible capacity of a software-defined and composable infrastructure.

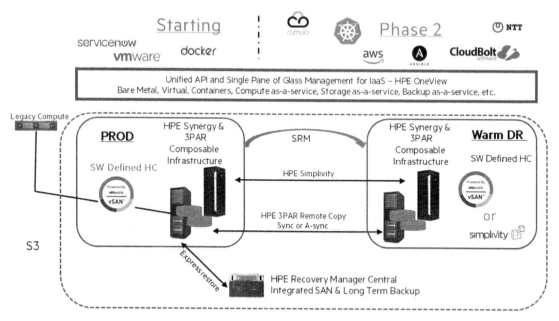

Figure 15-5 An example design showing common components

Figure 15-6 provides a snapshot of some of the key storage elements.

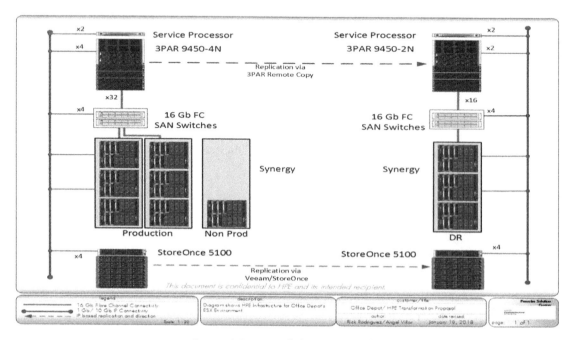

Figure 15-6 Partial design example

Hardware and software components

Table 15-2 shows the hardware and software components included in this design.

Table 15-2 Details of the example design

Product mapping	Functionality	Benefits
Infrastructure		
Synergy	Composable compute infrastructure	Unified API
3PAR Storage	Snapshot based replication. Storage for production and DR	Handles unpredictable workloads, application availability, and secure consolidation effectively
StoreOnce 5500	Local backups	Offers disk-based backup with deduplication for longer-term on-site data retention and off-site disaster recovery with best-in-class scalability and performance
VMware vRealize inbuilt IPAM	IP address management	Manage IP addresses using vRA inbuilt IPAM. Eliminate IP address conflicts and unused IPs
HPE Oneview	Bare metal provisioning, and infrastructure as a code	Unified API to manage infrastructure including compute, storage and network, and template-based profiles
Veeam data rotection	Backup software	
Utility		
vRealize automation	Hybrid Cloud\multi-cloud platform self-service portal, service catalogue, service broker, and resource management	Single platform to build a hybrid cloud with support to multiple heterogeneous resource providers(Compute\Storage\Network\Public Cloud)
vRealize Orchestration	Cloud orchestration, process automation, workflow engine	Automation and simplification of complex IT processes for IT, developer, and datacenter environment
vRealize Business	Metering and billing software for cloud platform	Optimize costs across private and public clouds, optimize resource utilization
Ansible (OD)	Infrastructure automation	Automation of hardware and infrastructure deployment
ServiceNow (OD)	User portal and approval workflow	Leverage ACME's investment in existing toolsets
Resource management and automation		
vRealize Orchestrator custom workflows along with Ansible	Software and Bare metal OS deployment, Patch management, audit and compliance management	Heterogeneous OS and virtualization platform support. Integration with Ansible for seamless integration with existing ACME environment

(Continued)

Table 15-2 Details of the example design—cont'd

Product mapping	Functionality	Benefits
vRealize operations manager	Business service-driven IT reporting software that provides resource, event, and response-time reporting across server, network and application environments	Capacity and resource monitoring to identify underutilized or overutilized resources in the cloud platform and optimize

HPE Datacenter Care Service

HPE Datacenter Care Service is HPE's most comprehensive support solution tailored to meet your specific data center support requirements. It offers a wide choice of proactive and reactive service levels to cover requirements ranging from the most basic to the most business-critical environments. HPE Datacenter Care Service is designed to scale to any size and type of data center environment while providing a single point of contact for all your support needs for HPE as well as selected multivendor products. The service is delivered under the governance of an assigned Hewlett Packard Enterprise support team that is familiar with your IT environment and understands how it enables your company's business. A mutually agreed-upon and executed Statement of Work (SOW) will detail the precise combination of reactive and proactive support features that will be provided under HPE Datacenter Care Service based upon your requirements.

Phased implementation approach

The design shown in Figure 15-5 can be implemented in phases, using virtual machines and containers as a precursor to provisioning IT-as-a-service using cloud-like consumption models:

1. Deliver VMs for Windows, Redhat, and SUSE Linux
2. Transfer operational control back to the IT department
3. Integrate on-premises backup
4. Replicate DR for 100% of production (65% of footprint)
5. Integrate reactive hardware support and pro-active Data Center Care for term of agreement

This approach allows for a 15% annual growth rate in demand for IT services while providing scope for future plans and integration of other roadmap items, such as:

- Integration with Docker (Future)
- Extension of services including MSSQL, Apache, and so on
- Extension to public cloud services

- Backup to the cloud
- DR in the cloud
- Extended operational support
- Migration services

Summary

This chapter shows how HPE enhances the IT consumption model with HPE GreenLake Consumption and Flex Capacity solutions. The various packages included in GreenLake portfolio of offerings provide a new, faster, and easier way for you to drive consumption-based discussions within your organization and provide exactly what is needed to meet your business goals. Now, you no longer have to choose between public cloud and on-premises IT. You can have the benefits of both, delivering what your business needs before they need it while maintaining full control within your own data center or on the edge. The sample architecture that is described shows how almost any type of HPE design can be implemented in the customer's datacenter. This in turn helps simplify the evolution to hybrid IT by delivering a cloud-like experience with flexible controls, whether the services are run on-prem or co-located. To quickly reap the benefits you are looking for, all you have to do is partner with a leader in consumption-based IT.

Why choose HPE?

Hewlett Packard Enterprise (HPE) is leading in consumption-based IT, with HPE GreenLake as its foundation. Built on industry-leading innovation, launched more than seven years ago, and with $1 billion contracted with customers worldwide and growing, HPE GreenLake is designed to meet customer initiatives with proven offerings suited to most every requirement. Consumption-based IT is an important enabler to any customer's IT strategy because not every application is able or ready to be ported immediately to the cloud; and this solution goes beyond just the infrastructure. It includes the additional services you need to deliver a broader and faster experience to the business. That is why you need to choose the right partner with the right expertise. As the next step in a long line of innovation, HPE GreenLake allows you to simply choose the outcome you want, and HPE Pointnext will do the rest. All under a single, pay-per-use model. To help you get up to speed faster, each solution includes an HPE Digital Learning subscription for e-learning content, to bring new skills to IT as needed.

CHAPTER 15
Consumption-Based IT Services

Additional resources

To learn more about fully designed, turnkey HPE GreenLake packages, use the following links:

- HPE GreenLake Edge Compute https://h20195.www2.hpe.com/V2/GetDocument.aspx?docname=A00026258ENW

- HPE GreenLake Backup https://h20195.www2.hpe.com/V2/GetDocument.aspx?docname=A00005054ENW

- HPE GreenLake for SAP HANA https://h20195.www2.hpe.com/V2/GetDocument.aspx?docname=A00036372ENW

- HPE GreenLake Database with EDB Postgres https://h20195.www2.hpe.com/V2/GetDocument.aspx?docname=A00036371ENW

- HPE GreenLake Big Data https://h20195.www2.hpe.com/V2/GetDocument.aspx?docname=A00016242ENW

16 HPE Secure Encryption

INTRODUCTION

Chapter 4 covered key Gen10 enhancements including "root of trust" security. This chapter will explore another key security feature offered ever since Gen8, HPE Secure Encryption. This solution secures data at rest on storage media such as Solid State Drives (SDD) and Hard Disk Drives (HDD.) Encryption on these devices greatly enhances the overall security of data.

HPE Secure Encryption is associated with using HPE Smart Array controllers and, optionally, an Enterprise Secure Key Manager (ESKM) hardware for secure key storage. The solution is applicable across many industries and easily scales with your business, ranging from a single server to enterprise-wide deployment of over 25,000 servers with ESKM. It addresses HDD and SSD connected to Smart Array controllers in ProLiant servers of Generations 8, 9, and 10.

HPE Secure Encryption is implemented in firmware and does not require any drive changes. The existing drives do not need to be swapped with Self-Encrypting Drives (SEDs). Therefore, it can be installed retroactively without making any changes to the existing storage media. It can also be uninstalled with minimal effort if a server needs to be repurposed without encryption. This is a very cost-effective and flexible solution that can be applied quickly to a small environment and subsequently grows to enterprise scale.

Why This Solution?

The solution was designed with the following considerations in mind.

Broad Encryption Coverage

Secure Encryption encrypts the data on both the attached drives and cache module of the HPE Smart Array controllers.

Secure Encryption Software supports all SSD and HDD for HPE ProLiant Gen8, Gen9, and 10 servers, and supported storage enclosures. This obviates the need for SEDs. Thus this solution is cheaper to implement, as it does not require changing drives to Self-Encrypting Drives (SED), which have 15%–20% price premium per drive over regular drives. SED SKUs are also very limited compared with non-Encrypted drives. By retaining the existing drives, Secure Encryption allows you to use and encrypt many more drive SKUs.

Flexibility

Secure Encryption is easy to install as it does not require new drives and backing up and restoring data. It is also more flexible if you wish to remove the encryption and repurpose the servers. Managing the Secure Encryption environment is simple and only requires a single license per server. ProLiant Gen10 features further enhance flexibility with Gen10 Redundant Arrays of Inexpensive Disk (RAID) controllers that simultaneously support RAID and non-RAID Logical Unit Numbers (LUN,) permitting you to run volumes that require encryption with others that do not require encryption on the same controller.

High availability and scalability

Secure Encryption easily scales with requirements from a single server to enterprise-wide deployment of over 25,000 servers using ESKM. ESKM manages all remote keys centrally and can be deployed in a clustered configuration for higher availability.

You can begin deployment in Local Mode, where keys are maintained at the Smart Array Controller. As deployment expands to enterprise levels such as hundreds of servers, an ESKM can be added for higher scalability. ESKMs can be deployed in clusters for higher availability.

Simplified deployment and management

The HPE Smart Storage Administrator (SSA) configures the features of Secure Encryption associated with the HPE Smart Array controllers. Only one HPE Secure Encryption license is required per server. SSA provides both graphical user interface (GUI) and Command Line Interface (CLI) to facilitate scripting. SSA was introduced mid-way in the Gen8 time frame. It replaces the Array Configuration Utility (ACU) completely in Gen9 and beyond. Most of the newer controller features, including HPE Secure Encryption, require SSA.

Regulatory compliance

HPE Secure Encryption helps enterprises comply with data privacy requirements associated with HIPAA, and Sarbanes-Oxley, PCI-DSS, HITECH and PII. It also supports all the major encryption algorithms (see Figure16-2 below).

Secure Encryption satisfies the cryptographic requirements established in the Federal Information Processing Standards(FIPS)140-2 by using National Institute of Standards and Technology (NIST)-approved algorithms in protecting both data and encryption keys. The following is a list of industries for which Secure Encryption is ideally suited:

- Government agencies and security contractors
- Financial, Healthcare, Insurance, Travel, e-Commerce (Web 2.0), and HR

- Level of assurance for drive returns or disposal (all industries)
- Compliments DMR, Decommissioning, and Relocations (across all industries)

Solution overview

The HPE Secure Encryption solution supports all drive models connected to Smart Array controllers in ProLiant Generations 8, 9, and 10. The solution only works on controllers in RAID mode (RAID volumes only). One license is required per server regardless of the number or type of drives and number or type of controllers on the server. Controllers may be external RAID controllers, supporting drives in attached external enclosures, such as D60x0, D3x00, and so on. One license can support hundreds of drives.

Note

- Gen8—controllers only support RAID mode
- Gen9—controllers support both RAID mode and HBA mode, but only one mode at a time with a boot required to change bode.
- Gen10—controllers can support both RAID mode and HBA mode simultaneously, on different volumes. Only RAID mode volumes can be encrypted with this solution. This enhances the flexibility of this solution since both RAID and non-RAID volumes can be supported simultaneously. Customers can thus simultaneously deploy LUNs requiring encryption and others that do not on the same controller.

Support for HPE Secure Encryption was first introduced midway in the life span of ProLiant Gen8 with the 12Gb SAS RAID controllers (12 Gb/s). The original Gen8 6Gb SAS RAID controllers does not support HPE Secure Encryption. Also, the Gen9 Smart Array B140i SW RAID, and Gen10 Smart Array S100I SW RAID do not support Secure Encryption. Here is a summary of the HPE Smart Array (SA) controllers that do support this encryption solution:

- Gen8 (Px3x)
- Gen9 (Px4x, Hx4x)
- Gen10 (All P-class and E-class controllers)

For the complete list, consult the QuickSPECS: http://h20195.www2.hpe.com/v2/redirect.aspx?/products/quickspecs/14623_div/14623_div.PDF

The solution supports ProLiant rack mount servers, C-class blades, Synergy compute modules, and attached drive enclosures. It allows for regulatory compliance and data protection and complements the HPE Pointnext services for data-wiping and defective media retention.

CHAPTER 16
HPE Secure Encryption

HPE Smart Array SR Secure Encryption is a FIPS 140-2 Level 1 validated enterprise-class encryption solution that complies with regulations for sensitive data. If you have HIPPA, Sarbanes-Oxley, PCI-DSS (Payment Card Industry), HITECH, and others, this solution may help with your compliance needs.

Encryption is often strictly regulated by local governments, so the below reminder is included, to ensure that local country laws be carefully considered before implementation. This is not an HPE Secure Encryption specific issue. It would apply to any product or service using encryption. The following disclaimer is an indication of the importance of checking on your specific compliance needs and local country laws before enabling the encryption:

The following paragraph is taken from the QuickSpecs document:

> HPE Special Reminder: Before enabling encryption on the Smart Array controller module on this system, you must ensure that your intended use of the encryption complies with relevant local laws, regulations and policies, and approvals or licenses must be obtained if applicable. For any compliance issues arising from your operation/usage of encryption within the Smart Array controller module which violates the above mentioned requirement, you shall bear all the liabilities wholly and solely. HPE will not be responsible for any related liabilities.

Key management

Encryption is accomplished using keys. Keys are long sequences of numbers, mathematically constructed to transform the original plain-text data into unreadable cypher text. Keys are intended to make recovery of the original plain-text difficult. As there are potentially many keys to manage, Secure Encryption can manage the keys in either Local Mode or Remote Mode. In Local Mode, keys are kept at the RAID controller. With Remote Mode, keys are stored on and managed at an ESKM device.

Local mode: Keys are maintained and managed at the RAID controller. This is designed for use in a small environment, or the initial phase of deployment.

Remote mode: This requires Integrated Lights Out (iLO) Advanced or Scale Out editions v1.40 or later and ESKM 3.1 or later release. This is designed for use with an enterprise scale environment and can support up to 25,000 servers and millions of keys. As noted above, ESKMs are often deployed as clusters for both scalability and high availability. Remote Mode currently requires the use of HPE's ESKM. Although the HPE ESKM supports the Key Management interoperability Protocol (KMIP), the ProLiant server firmware currently does not. Remote Mode requires ESKM. HPE's ESKM hardware and software is now sold through Micro Focus. See **http://hpe.com/software/eskm**

Encryption algorithms

Secure Encryption is based on standards outlined in FIPS 140-2. Controllers using Secure Encryption implement both physical security and cryptographic methods in protecting data-at-rest. Specifically, Secure Encryption satisfies the cryptographic requirements established in FIPS 140-2 by using NIST-approved algorithms in protecting both data and encryption keys as shown in Table 16-1.

Table 16-1 Encryption algorithms

Algorithm	Description
XTS-AES 256-bit	The XTS algorithm is used to encrypt data on the drive platter as described in NIST special publication SP 800-38E.
AES-ECB The AES	The AES algorithm is used to perform symmetric key encryption.
SHA-256	The SHA secure hashing algorithms are described in FIPS 180-4.
HMAC	The HMAC algorithm is described in the FIPS 198-1 standard.
PBKDF2	The PBKDF2 algorithm derives cryptographic keying material from user-provided passwords. The algorithm is described in NIST special publication SP 800-132.
DRBG	An implementation of the SP800-90A algorithm is used to produce random bit sequences.

Federal information processing standards (FIPS) compliance

- The ESKM is FIPS 140-2 Level-2 **validated**, certificate #1922
- The HPE Smart Array Gen8 Px3x family of controllers is FIPS 140-2 Level-2 **validated**; certificate #2375
- The HPE Smart Array Gen9 Px4x controllers P244br, P246br, P440, P441, and P741m are FIPS 140-2 Level-1 **validated;** certificate #2506
- FIPS 140-2 Implementation Under Test List for Gen10 Smart Array Performance RAID P-class Controllers

Enterprise Secure Key Manager (ESKM)

With Remote Mode, HPE Secure Encryption uses ESKM to manage all the keys. This is intended for an enterprise environment with up to 25,000 servers and possibly millions of keys and drives. Communication between the server and ESKM is proprietary and the only ESKM currently supported is the HPE ESKM. HPE ESKM solution **does** support KMIP and thus the HPE ESKM solution can manage keys from **non-HPE** environments that support KMIP.

- ESKM is a product built as a hardened appliance. It has been validated as a complete solution under FIPS 140-2 Level 2 requirements for Cryptographic Modules. Operating in FIPS mode can be configured by the administrator. When in FIPS mode, only use of FIPS-approved algorithms for cryptographic operations is allowed. Nonapproved TLS/SSL cipher suites will be disabled and only FIPS-approved keys can be created. The ESKM will also disable global keys, FTP, LDAP, and SSL 3.0. HPE recommends using FIPS mode.
- The ESKM is based on standard ProLiant hardware and can be clustered for scalability and availability. Up to eight nodes can comprise a cluster, for shared services in large enterprises spanning multiple datacenters and geographies.

- Remote configuration and management is available through a secure Web-based GUI and a command-line interface (CLI).

- Management security is provided with SSL communications, password-based authentication, fine-grained identity-based administrator privileges, audit logging, and multiple credentials for critical actions

- Logging and monitoring: supports logging of all events, external Syslog/SIEM servers, SNMP v1/2/3 traps, and FIPS/Health check status servers.

- Remote configuration and management: is available through a secure Web-based GUI and a command-line interface (CLI)

- Management security: is provided with SSL communications, password-based authentication, fine-grained, identity-based administrator privileges, audit logging, and multiple credentials for critical actions

- Logging and monitoring: supports logging of all events, external Syslog/SIEM servers, SNMP v1/2/3 traps, and FIPS/Health check status servers.

- Administrators should consult the ESKM Key Protection Best Practices document at https://4b0e0ccff07a2960f53e-707fda739cd414d8753e03d02c531a72.ssl.cf5.rackcdn.com/wp-content/uploads/2016/10/4AA2-1403ENW.pdf?v=20

Licensing

One server license per server is required for all the controllers and drives (internal and external) attached to the controllers in the server. Customers must secure the license below per server enabled for HPE Secure Encryption..

HPE Smart Array SR Secure Encryption (Data at Rest Encryption/per Server Entitlement) ELTU Q2F26AAE

If also using Remote Mode key management, the following licenses are also needed:

- A license for Integrated Lights Out Advanced or Scale Out editions v1.40 or later, and requires HPE Enterprise Secure Key Manager (ESKM) 3.1 or later release (typically sold in a clustered pair.), and a

- A client license per HPE ProLiant server connected to the ESKM. See HPE ESKM information for more details. http://hpe.com/software/eskm

HP ESKM Additional Client License   BB741AA

Note **one** client license is included with each ESKM (AJ575A)

The versions above are for iLO 4 and are minimum versions. It's highly recommended that customers use the most (or at least a more) recent version. iLO5 (Gen10 servers) is supported from the initial release.

Considerations

- Examine each ProLiant servers to determine the supporting controllers and the possible need to change/add controllers. This is especially pertinent in Gen8 systems, where the embedded controller, P420i, predated the HPE Secure Encryption offering. Some controller configurations/additions may be needed that would also require validating slot availability.

- Review the list of countries where deployment is planned to ensure compliance with all local encryption regulations and laws.

- When considering an HPE Secure Encryption, consult the HPE Smart Array SR Secure Encryption Installation and User Guide, https://support.hpe.com/hpsc/doc/public/display?docId=a00018950en_us for the configuration steps.

- Decide on the scope of the deployment Local Mode (small scale) versus Remote Mode (large scale, enterprise-wide. You can start with Local Mode and later as you grow migrate to Remote Mode.

- Remote Mode would require the purchase of ESKM appliances and cluster topologies (how many nodes, where). Determine the number of additional ESKM client licenses required.

- Consult the Smart Storage Administration User Guide https://support.hpe.com/hpsc/doc/public/display?docId=c03909334

- For large environments, use the CLIs and scripting to automate.

- Determine security administration roles.

Competitive view

Many competitors use Self-Encrypted Drives (SEDs). Table 16-2 below lists the advantages of HPE Secure encryption over the SED alternative.

Table 16-2 HPE Secure Encryption versus SED-Based Solutions

	HPE Secure Encryption	**SED based Solutions**
Drive Collection	All drives supported on HPE ProLiant Servers Gen8, 9, and 10	Very limited drive models and capacities
Cost	License cost (fixed list cost per server)	15%–20% price premium per drive
Encryption coverage	Drives + controller Write Cache	Drives only

System Management

In Local Mode, this solution is managed primarily with the Smart Storage Administrator (SSA). See the HPE Smart Array SR Secure Encryption Installation and User Guide (Encryption Manager) https://support.hpe.com/hpsc/doc/public/display?docId=a00018950en_us

If using Remote Mode, then the ESKM is configured and managed via the ESKM's software. Connection to the ESKM is through the server's iLO 4 or 5.

I. **Log in to the ESKM**

II. **Create initial user accounts**.

   a. Create an account called **DeployUser**.

   b. Create an account called **MSRUser**.

III. **Create a group**.

IV. **Assign the user account for hosting Master Encryption Keys to the group created in step 3**.

V. **Create a Master Encryption Key to be used by the controller**.

Be sure to set the owner of the key to the user account created to host the Master Encryption Key created in Step 2b.

VI. **Place the Master Encryption Key in the group created in step 3**.

See the Secure Encryption Installation and User Guide https://support.hpe.com/hpsc/doc/public/display?docId=a00018950en_us

Related solutions

HPE has invested heavily in multiple security solution that are documented and verified by third parties. This has been a steady effort throughout each ProLiant generation, but especially so in the latest generation, Gen10. HPE security solutions are across all our product families and are comprehensive across the entire lifecycle of our products. In addition to product enhancements, we have added many Point Next services for security. See http://www.hpe.com/security

Summary

The HPE Secure Encryption solution covered in this chapter supports multiple ProLiant server and blades generations (Generations 8, 9, and 10) and provides many benefits, including:

- Supports a very wide set of drive models and capacities. It is not limited to the small set of SED SKUs.
- Using standard drives also enables ease of migration from a nonencrypted to an encrypted environment. There is no need to change drives and for backup/restore (data migration).

- Encrypts the drives and the controller's cache.
- Lower cost—based on lower cost of standard drives versus SEDs.
- Flexibility—supports both small tactical environments (Local Mode) and enterprise scale environments (Remote Mode with ESKM)
- Supports the last three generations of ProLiant systems.
- Remote Mode uses ESKMs originally designed by Atalla. Atalla has been part of the Tandem/Compaq/HPE family for over 30 years and a major producer of Secure Key storage systems.

Additional resources

HPE Secure Encryption Web site
https://www.hpe.com/us/en/product-catalog/detail/pip.6532260.html

HPE Secure Encryption QuickSpecs
https://h20195.www2.hpe.com/V2/GetDocument.aspx?docname=c04318075

HPE Smart Array Secure Encryption Installation and User Guide
https://support.hpe.com/hpsc/doc/public/display?docId=a00018950en_us

Enterprise Secure Key Manager (ESKM)
http://hpe.com/software/eskm

HPE Smart Storage Administration User Guide
https://support.hpe.com/hpsc/doc/public/display?docId=c03909334

HPE Smart Array SR Gen10 Configuration Guide
https://support.hpe.com/hpsc/doc/public/display?docId=a00018944en_us

HPE ESKM – Key Protection Best Practices
https://4b0e0ccff07a2960f53e-707fda739cd414d8753e03d02c531a72.ssl.cf5.rackcdn.com/wp-content/uploads/2016/10/4AA2-1403ENW.pdf?v=20

HPE Security Web Site
http://www.hpe.com/security

17 Blockchain Demystified

INTRODUCTION

Every 10 to 15 years, there is a major technological revolution. Mainframes, personal computers, smartphones, cloud, and artificial intelligence have all transformed the way we live and work. Blockchain too has the potential to revolutionize many industries and improve lives. The technology industry will have to once again reengineer solutions and approaches to support this potentially rich but still evolving technology.

Blockchain's first applications targeted the financial services industry. Transportation, logistics, healthcare, and public records applications are in proof-of-concept and active development today. The technology evolution will naturally move from the first implementation, which was public and not regulated, to a set of private implementations that will provide the regulations and controls which are crucial for wide-scale adoption.

What if every kind of asset from money to music could be stored, moved, transacted, exchanged, and managed, all without powerful intermediaries? Marc Andreessen, a venture capitalist who invested in some of the most transformative companies in recent history including Facebook and Twitter, says "Blockchain's distributed consensus model is the most important invention since the internet itself."[1]

In order to better understand any complex technology like blockchain, it helps to understand the history, the reason for its development, and some of the key components and concepts that make up this potentially disruptive technology.

To some the 2007–2008 financial crisis was seen as a fundamental breakdown of the global banking system. Government insured banks around the world seeking higher returns for their shareholders sought these returns through complex derivatives tied to the housing market. Ultimately, these new, less-regulated, instruments proved to be much more risky than previously known. This resulted in a veritable "run on the banking system" where investors and depositors around the world pulled money out of these institutions that put the stability of the bank and the broader global banking system at risk of collapse.

Governments around the world, fearing a devastating financial collapse, allocated public funds to back stop these institutions. These funds in some cases came from central treasuries issuing government-backed debt and effectively printing new money. These dramatic actions were seen by some as a breakdown of capitalism as private institutions were protected with tax payer funds, resulting in the devaluation of the currency and an increase in the national debt.

---

[1] Blockchain technology is changing the approach to solve business problems and streamline operations.

CHAPTER 17
Blockchain Demystified

It is believed that a person or persons known under the pseudonym Satoshi Nakamoto saw these historic events as concerning and imagined a currency alternative not tied to a central bank and less impacted by major economic or political events. This led to the development of a decentralized peer-to-peer platform, blockchain, and its first application, bitcoin, in 2009, the world's first cryptocurrency.

The potential to streamline critical processes with a high degree of security across broad industry groups is the reason blockchain is widely considered the next disruptive technology. Eliminating intermediaries, completely reengineering the concept of contracts, will see traditional businesses disappearing while new ones will be created. The need for standards created three major consortiums backed by technology giants and blue chip companies. They are Corda R3, which is focused on financial services applications, Hyperledger that is focused on logistics, transportation, and general industrial applications and Ethereum. These are discussed in more detail later in the chapter.

We will cover many topics related to blockchain in this chapter including the following:

- Fundamentals of blockchain: this section describes the various technologies that enable current blockchain frameworks and applications.

- Blockchain public, private, and consortium: this section provides a basic description of each.

- Blockchain use cases: the various use cases of the technology.

- Challenges: the various headwinds impacting the adoption of the technology.

- HPE solutions: this section describes the HPE approach to holistic blockchain solution for our clients.

Blockchain fundamentals

In its simplest form, Blockchain is a digital ledger of transactions that has applications in virtually every industry. Electronic transactions of virtually any type can be validated using blockchain technology. A "block" is a container data structure that contains a header and a list of records or transactions. A blockchain is updated when a transaction is verified, added to a block, and that block is added to the existing "chain." The simplified sequence of events has a flow described in Figure 17-1.

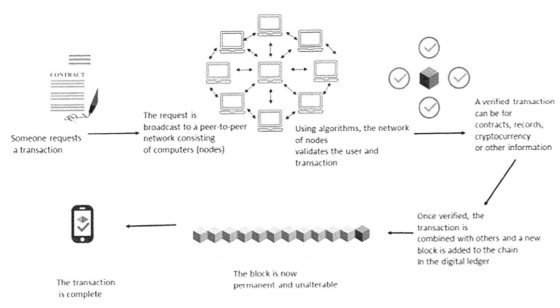

Figure 17-1 Simplified Blockchain flow

The following describes the sequence of steps shown in Figure 17-1:

1. A transaction of some type is requested
2. A peer-to-peer network with a group of nodes on it broadcasts the request
3. The network of nodes validates the transaction based on the user's status using know algorithms
4. The transaction is complete and combined with other transactions that creates a new block for the ledger
5. The new block is added to the chain of existing blocks making it permanent and unalterable

Cryptography and crypto currencies

Cryptography in its basic form is using mathematics to encrypt and decrypt data so that it can either be stored or send to someone. Leveraging encryption only the intended recipient can read it. We take plain text unencrypted data, and encrypt it using a cipher, a mathematical algorithm used to securely encrypt and decrypt data, to produce ciphertext that is unreadable encrypted data. Conventional cryptography will leverage the same key to encrypt and decrypt the data. This is called symmetric-key cryptography. Figure 17-2 depicts the process of encryption:

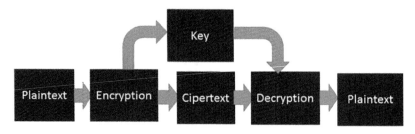

Figure 17-2 The encryption process
Source: https://chrispacia.files.wordpress.com/2013/09/digital-signature.jpg

Cryptography is essential to Bitcoin and what makes Bitcoin possible. Public-key cryptography has a second benefit beyond just the encryption and decryption of data. It can be used to create something called a digital signature that can be used to simultaneously provide authentication, data integrity, and nonrepudiation, all of which are critical to Bitcoin's operation.

A digital signature is generated by combining a user's private key with the data that he wishes to sign in a mathematical algorithm. Once the data is signed, the corresponding public key can be used to verify that the signature is valid. Figure 17-3 depicts the signature process:

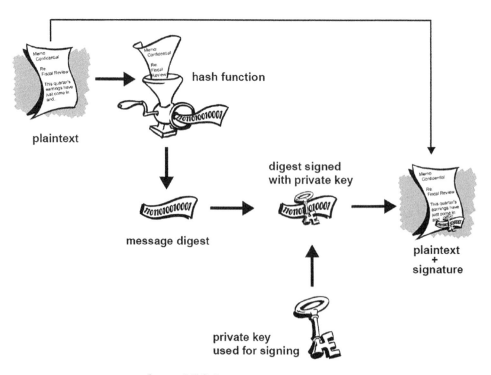

Figure 17-3 Bitcoin signature process
Source: https://chrispacia.files.wordpress.com/2013/09/digital-signature.jpg

A **cryptocurrency** is classified as a digital asset that is designed to work as a medium of exchange. It uses cryptography to secure the transactions, control the creation of additional units, and to verify the transfer of assets. Cryptocurrencies can be classified as a subset of digital currencies and are also classified as a subset of alterative currencies and virtual currencies. The most prevalent cryptocurrency today is Bitcoin and several other cryptocurrencies such as Ethereum and Litecoin had emerged as a Bitcoin alternative.

Bitcoin is a digital cryptocurrency combination of BitTorrent technology (peer-to-peer file sharing) and public key cryptography. One of the core functions is to handle the double-spend problem. For example, digital assets can be copied. Digital cash, like an image attached to an email, can be copied many times. A centralized third party is required to issue and reconcile digital cash transactions to prevent duplicates. The implication is that **any** online transaction can be decentralized and conducted in a peer-to-peer trustless manner without any controlling authority.

Distributed Ledger: A distributed ledger is aledger of any transactions or contracts maintained in decentralized form across different locations and people, eliminating the need of a central authority. The intent is to keep a check against manipulation.

A **Node:** Any computer that connects to the blockchain network and that fully enforces all the rules of a specific blockchain implementation is called anode.

Blockchain frameworks

There are **three major frameworks** that have emerged leveraging the blockchain technology: Hyperledger Fabric, R3 Corda, and Ethereum. Each have different visions in mind with respect to industry applications. Where Hyperledger Fabric is an IBM-led codebase modular blockchain platform that has several industry applications such as supply chain, FSI, banking, and healthcare the R3 Corda industry applications focus squarely on financial services. Ethereum is a generic blockchain platform independent of any specific field.

Hyperledger fabric: The Hyperledger fabric is a leading open source blockchain fabric built for enterprise. With IBM as a primary contributor, it is the most prevalent Hyperdeger project hosted by the Linux foundation. It is designed for developing Blockchain applications with a modular architecture so components such as consensus and membership services can be plug and play.

R3 Corda: This is a distributed ledger designed for Financial Services. It focuses on recording and managing financial assets. One of the attributes of Corda that makes is specifically suitable for the FSI industry is that has no unnecessary sharing of data,and, only the parties with a legitimate use can see the data whiten an agreement and supports a variety of consensus mechanisms. It is built on industry standard tools and as it stands corda has no native cryptocurrency applications. Ultimately, corda will support a variety of consensus mechanisms, digital tokens might be present, or allowed to be used in its ledger system.

Ethereum: This open source, public, distributed blockchain-based distributed computing platform and operating system with smart contract functionality was first released in July 2015. It allows developers to build and deploy decentralized application with cryptocurrency being just one example of an application.

Blockchain: Public, private, and consortium

As businesses look to take advantage of blockchain technology, they look to find designs that complement their existing business. One of the first decisions is whether to deploy on a public, private, or consortium-based distributed ledger. So the first question is, why so many choices? Bitcoin is deployed using a public blockchain, and while there have been plenty of headlines, Bitcoin remains a very secure form of currency.

The public Bitcoin blockchain was designed to remove the middleman in a transaction. It accomplishes this by creating a digital record or "block" of peer-to-peer transactions. Each transaction needs to be synced and verified with every node associated with the blockchain before it is written to the ledger. This is done serially with the next transaction waiting in queue. The number of nodes is virtually limitless as anyone with a computer and access to the Internet can become a node and sync with the entire blockchain. Any valid node on the network can read and send transactions. Anyone can participate in the validation and consensus process to determine which blocks get added to the chain. This consensus process includes financial payments to node operators for providing the validation services that drive the blockchain engine. While the design of this public system is highly secure and stable because of the number of nodes and complexity of the validation process, it also makes it very slow and highly wasteful considering the massive amount of power and compute resources needed to validate each transaction.

This public model is most appropriate when the exchange needs to be decentralized, usage needs to be anonymous, and transactions need to be fully transparent. Costs are higher and processing may be less efficient than other systems, but the cost benefits are significant over legacy transaction platforms that use intermediaries to validate the transaction.

A private or a consortium model as opposed to the public platform that Bitcoin uses, allows companies to retain privacy and control while still reducing the time and cost of many of today's inefficient transaction systems. As existing businesses such as financial institutions start looking at the advantages that blockchain technologies could offer they will logically come-up with architectures that complement their existing business operations. Thereby the private or consortium model will logically be more popular with these types of businesses.

A consortium model for a blockchain exists where the consensus process is managed by a limited and pre-defined set of nodes. For example, imagine a set of five banks who want to process transactions among themselves. There would be six nodes on the network. One from each bank and a regulator node. Each node would need to validate every block. The right to read transactions could be public or limited to specific participants based on the nature of the transactions being processed and the competitive dynamics among the banks.

A fully Private blockchain is one where write permissions are controlled by one organization. Read permissions could be made public or restricted as needed. Potential applications for a private system could include accounting, administrative, and auditing functions within a single company.

In some cases, the private model could result in lower overall costs and faster transaction times than a public blockchain as the number of nodes can be reduced in this model.

Some purists see the growth of private blockchains as a protectionist attempt by large financial institutions to retain control of the financial system. Purists continue to advocate for a fully decentralized public blockchain. Most blockchain advocates believe that both have their inherent advantages and disadvantages, but in general, the benefits of improved efficiency and lower cost are hard to argue with.

Blockchain: Use cases

Figure 17-4 shows that the potential applications for blockchain technology are expansive, but here we will focus on the main areas of interest within the Financial Services Sector.

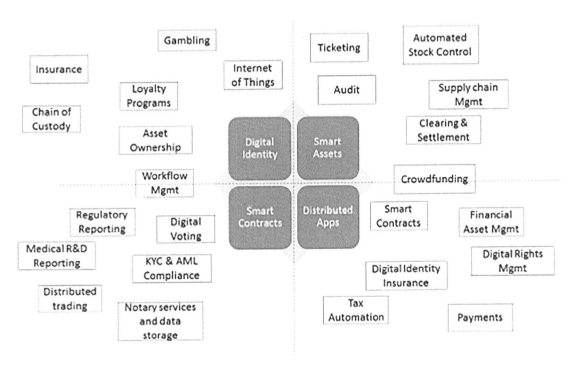

Figure 17-4 Blockchain applications

Blockchain activity within financial services is largely focused on the consortia and networks listed below. Figure 17-5 provides an overview of where key participants are focusing.

CHAPTER 17
Blockchain Demystified

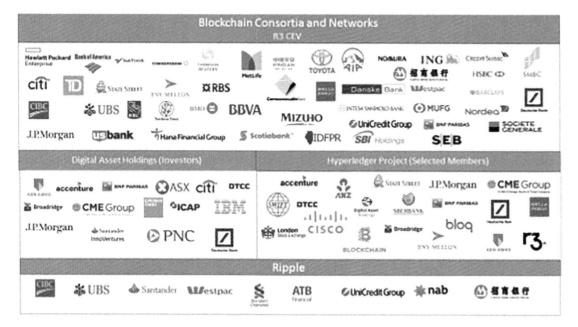

Figure 17-5 Blockchain participants

Source: BI Intelligence

Smart contracts

Smart contract is a term used to describe computer program code that is capable of facilitating, executing, and enforcing the negotiation or performance of an agreement (that is, a contract). The entire process is automated and can act as a complement or replacement of a legal contract. This is not a mere digitation of the legal contract. Instead parties to a contract create and agree to a simple set of requirements and instructions that are required for establishing and governing an agreement. While complex agreements may not be easily managed by code; simple agreements like rentals and asset sales could be dramatically streamlined by leveraging blockchain technology. Figure 17-6 depicts the smart contact flow.

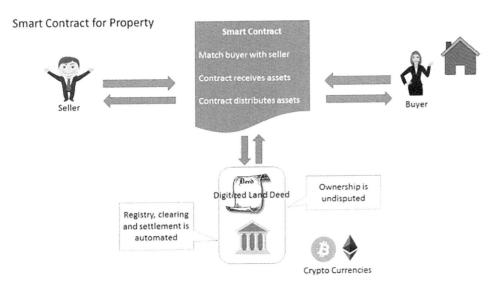

Figure 17-6 Smart contract

Source: https://blockgeeks.com/

Smart assets

Typical asset sales can benefit from a blockchain by simply implementing an immutable chain of transactions that includes date, time stamps, and terms of sale. But expand this into a complex supply chain, such as automobiles or beef, and you create a real-time information system providing critical data to every participant in the value chain. The key will be an enriched dataset where components or assets in a supply chain will not only include serial number and price but other relevant information like manufacturer, origination, and destination details; how it relates to other assets in the supply chain; transportation; entry and exit port details; the components and component value that are included in the asset; tax and government regulatory information and others. This enriched data can result in a safer more stable supply chain along with creating a competitive differentiator to any firm who can exploit its inherent value. Figure 17-7 illustrates how the concept of Smart Assets can be leveraged in the beef industry.

CHAPTER 17
Blockchain Demystified

Figure 17-7 Supply chain

Source: http://www.oliverwyman.com/our-expertise/insights/2017/jun/blockchain-the-backbone-of-digital-supply-chains.html

Clearing and Settlement

Clearing and settlement applications appear to be the most active areas for blockchain technology. Experts believe that the traditional three-day clearing and settlement process associated with many financial transactions cost financial institutions around $20 billion per year in overhead associated with custodial services, central counterparty systems, collateral management, and end of day reconciliations processes. Figure 17-8 shows the differences of a centralized versus a distributed ledger.

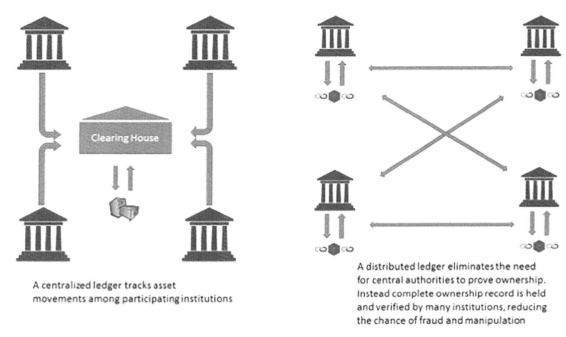

Figure 17-8 Ledgers

Source: FT Research

Payments

The possibility of true peer-to-peer payment systems is enticing financial disruptors and traditional financial intermediaries alike. High fees, high volume, and relatively low risk have created a $1.8 trillion industry that is clearly in the cross hairs of many blockchain enthusiasts. R3, the consortium, backed by 42 of the world's largest banks are developing a bank-to-bank global transaction system based on the Ethereum general ledger platform.

CHAPTER 17
Blockchain Demystified

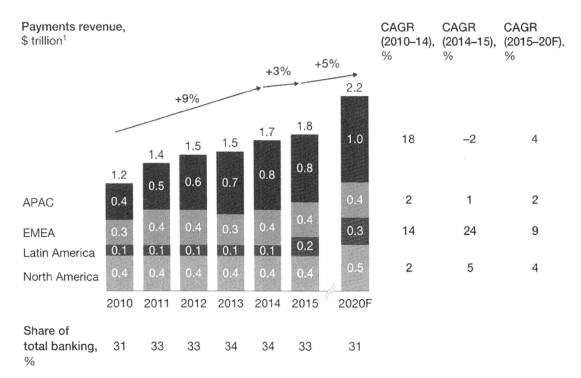

Figure 17-9 Global payments industry

Source: https://www.mckinsey.com/industries/financial-services/our-insights/a-mixed-year-for-the-global-payments-industry

An example of a Distributed Ledger Payment System consists of validating nodes, a certificate authority, and client applications. Validating nodes are responsible for endorsing and maintaining the ledger state by confirming transactions. The certificate authority distributes, manages, and revokes user privileges. Client applications send transactions to nodes as shown in Figure 17-10.

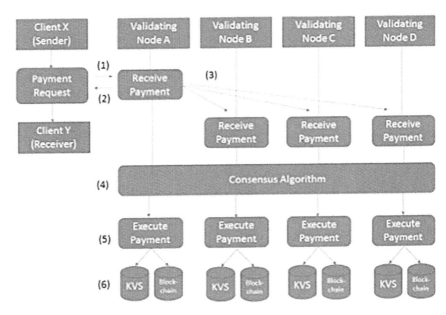

Figure 17-10 Distributed Ledger Payment System

(1) Client X Sends a payment with a digital signature to Client Y.

(2) Validating Node A acknowledges receipt of the payment.

(3) Validating Node A broadcasts the ordered payments to the other validating nodes.

(4) Validating nodes verify the payments by reaching a consensus based on a predetermine algorithm that includes validation of transaction specific details like digital signatures and transaction serial numbers.

(5) Validating nodes execute the payment using a smart contract after receiving a commit message from the minimum number of nodes required to achieve consensus.

(6) Validating nodes record the updated status and create a new block on the ledger.

* KVS = Key Value Store

Source: https://www.ecb.europa.eu/pub/pdf/other/ecb.stella_project_report_september_2017.pdf

Digital identity

As the sharing economy grows and the promise of IoT creates a world where autonomous devices interact with each other on our behalf a system of authentication is essential as more and more transactions can be associated with individuals and businesses. Our identities will need to be recorded on a general ledge along with all the devices enabled to operate within a machine-to-machine economy.

The Electronic Health Record (EHR) is a digital version of a patient's medical history maintained by a healthcare provider over the course of their visits. The record includes patient information on demographics, diagnosis, vital signs, past medical history, progress over time, lab tests, and so forth. Some major advantages offered by an EHR system include accurate and up-to-date patient information, reduced cost, quick access to patient related data, reduced medical errors, increased patient participation, and improved efficiency of healthcare providers.

The biggest challenge that is being faced by health-care systems throughout the world is how to share medical data with known and unknown stakeholders for various purposes while ensuring data integrity and protecting individual privacy. Although data standards have improved, each EHR stores data using different workflows, so questions of who recorded what and when is not always clear reducing confidence levels for medical professionals and their patients when assessing critical decisions.

Several organizations and consortia have kicked off programs to leverage blockchain technology to tackle this problem. MIT's MedRec project supports the use of Ethereum smart contracts to automate and track certain record state alterations (such as a change in viewership rights, or the birth of a new record in the system). The smart contract logs patient/provider relationships that associate a medical record with viewing permissions and data retrieval details for execution on external databases. They include a cryptographic hash of the record to ensure against tampering, thus guaranteeing data integrity. Providers can add a new record associated with a particular patient, and patients can authorize sharing of records among providers. In each case, the parties are automatically notified of any state change to approve or reject the proposed change. Keeping participants informed and engaged in the evolution of their records.

Blockchain challenges

Assessing the risks to and from Blockchain is a challenge in and of itself. Indeed, this section is perhaps the most difficult to write since it offers speculative views on how this new technology will be embraced and used.

What are the headwinds facing this new Technology?

As we have discussed, many of the most compelling use cases for Blockchain are in the financial services and banking sector. Payments and trade clearing are two of the ripest areas. For the industry and the public to embrace a change, they must be greeted by a faster, more secure and less-expensive solution. The general nature of a public blockchain, while eliminating the intermediary introduces an ever more lengthy validation process that grows more costly and slow as the network grows. In addition, as the number of nodes grow on a public blockchain the incremental profit in competing to solve the block algorithm goes down as each node is added. Average transactions fees have increased to $99 while the time to validate a transaction is now around 45 minutes. Figure 17-11 shows an example Bitcoin chart.

Figure 17-11 Bitcoin chart
Source: Coindesk - https://www.coindesk.com/price/

Figure 17-12 shows an example of cost-per-transaction.

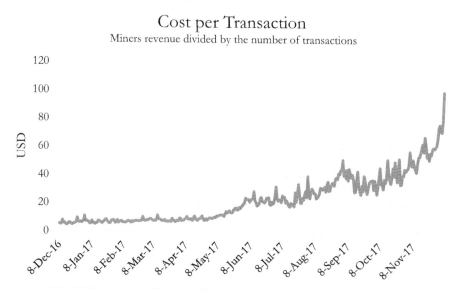

Figure 17-12 Cost per transaction

*https://blockchain.info/charts/cost-per-transaction

Figure 17-13 shows Bitcoin confirmation time.

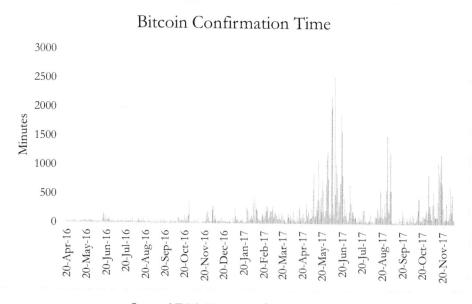

Figure 17-13 Bitcoin confirmation time

* https://blockchain.info/charts/avg-confirmation-time?timespan=all

These factors may slow adoption because the same factors that could eliminate a costly intermediary could also make the system slow and potentially more expensive if validators see a reduction in the commercial benefits of participating in the process.

When Smart Contracts are not so smart

Recently, the Ethereum public blockchain was rocked by a participant who exploited a flaw in a "Smart Contract" that allowed the thief to syphon off millions of the crypto currency while the Ethereum community struggled to reach consensus of what to do. While distributed ledge technologies like the one employed in Ethereum are encrypted and extremely difficult to hack the contracts that are created to govern the commercial agreements are nothing more than programs that manage the transaction details. These programs can be hacked. In the case of the Ethereum breach, the hacker exploited a hole in the contract that allowed the attacker(s) to move about $45 million worth of the currency into their own account. The nature of the public Ethereum distributed ledger required that all participants in this smart contract move to a new version of the ledger effectively requiring a new fork off the current version of the Ethereum code. This required all participants to agree to move. This is one of the main drawbacks of a public blockchain with no central authority to govern how the community responds to a disruptive event.

When the distributed ledger is public

The core trade off of the distributed ledger and blockchain is the fact that all transitions are public. While there is some anonymity to the ledger, some participants may fear that a competitor can use this public information to their advantage. This will surely push some to focus more on private or consortium structures, which could reduce the benefits to participants as companies that are likely to get disrupted by the technology may limit innovation to avoid the potential for reduced revenues or lower margins.

Lack of common architectures

Finally, one key factor in the success of any disruptive technology is the presence of a well understood and broadly adopted set of standards. What drove the explosive growth of the Internet was a proven network architecture and transport protocol coupled with the common Internet programming language of HTML. This allowed individuals, schools, government agencies, and businesses to starting leveraging and investing in the technology knowing they did not have to choose between competing platforms.

At this time there is no proven, well-integrated set of technologies that can be leveraged for a private or consortium use case or business model. Both R3 and HyperLedger have demonstrated some promising solutions, but none are being leveraged at scale in production. Thus, the question of whether blockchains can work in real-life scenarios is still an open question.

Hewlett Packard Enterprise solutions

Enterprises evaluating blockchain solutions are finding that generic infrastructure and public cloud environments cannot support their unique use cases and business requirements. To better address the growing demand, HPE is expanding its service offering to include Mission Critical Distributed Ledger Technology Solution that offers availability and fault protection for enterprise-grade applications.

HPE provides consulting services to help customers determine the best choice for their business. Some solutions require a single tool, but most require a collection of integrated components and a design that embraces legacy processes and data. Further, HPE works with certified partners to provide additional options for delivery of this technology into the market providing development, customization, and lifecycle management. Some examples of these use cases and services are outlined below especially with regard to HPE and Distributed Ledger Technology in the Financial Services Industry (FSI).

Distributed ledger use cases for financial services

Many of today's blockchain use cases are financial and involve the use of crypto-assets either within the deployment design or as the sole purpose of the deployment like Bitcoin or Ethereum's Ether Coin.

The distributed ledger system in FSI extends the potential uses beyond crypto currency. Financial Industry demands and concerns will drive mixed deployments of the technology.

Distributed Ledger Technology in FSI

- Clearing
- Settlement
- Trade finance
- Mortgages
- Title management
- Physical asset registration
- Trade execution and settlement
- Asset exchange
- Cash reserve management
- Supply chain management

HPE Mission-Critical Distributed Ledger Technology (MCDLT)

As discussed earlier, R3 Corda is a transactional distributed ledger solution that can deliver enterprise-grade services with significant benefits:

- Standard DLT functionality
- Immutable ledger
- Distributed (secured visibility)
- Certified (smart contract, legally binding)
- Open-source design
- User-designed transaction flow logic
- On-premise or Cloud deployment

R3 Corda smart contract key concepts

Figure 17-14 shows the fundamentals of smart contracts.

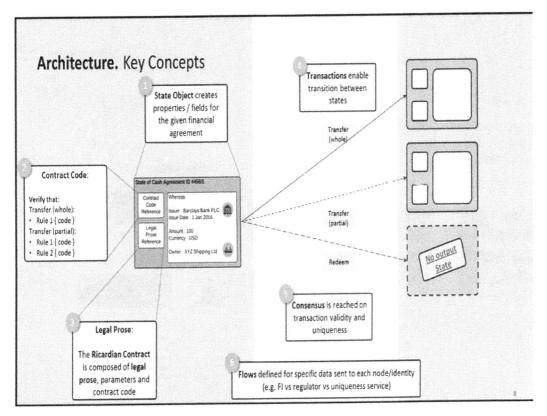

Figure 17-14 Smart contracts

*Content extracted from R3 Corda Early Access program

CHAPTER 17
Blockchain Demystified

HPE adds enterprise-grade enhancements to R3 Corda using the advanced features of the HPE NonStop Server including:

- Linear scale-out for enterprise application support
- Always-on services for the execution environment
- Hardened enterprise-grade relational database—SQL/MX product
- Guaranteed message delivery (Figure 17-15)

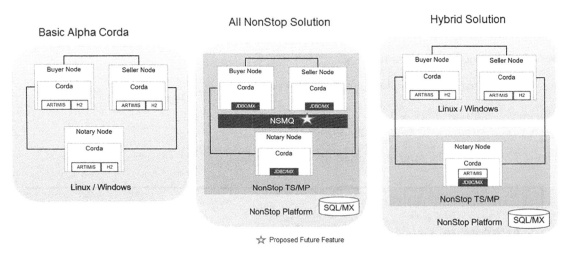

Figure 17-15 HPE NonStop solution

Enterprise data integration

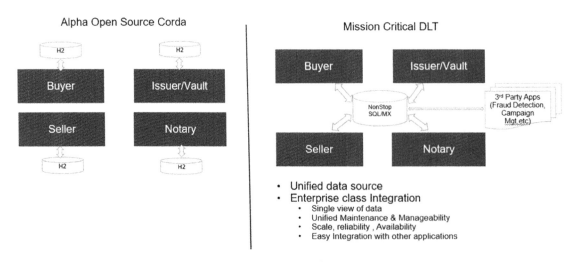

Figure 17-16 Corda SQL

This Mission-Critical DLT adds significant value to the R3 Corda's open-source platform by providing enterprise-grade transactional DLTs. HPE expects the Financial Services firms who need heightened availability and performance to deploy R3 with HPE's mission-critical DLT for many use cases (Figure 17-16). Here are some examples:

Settlement

The confirmation of a financial transaction in the capital market generally requires a consensus of certified central intermediary like an Exchange. The promise of a smart contract DLT services is that it provides certified, immutable proof that can be used to confirm settlement among financial parties. Key features of a mature settlement process includes:

- Privacy (cryptographically strong data at rest)
- Immutability
- Transactional access authentication

HPE Mission-critical blockchain addresses all of the above needs.

Asset registration

Asset management requires coordination among management of position, established value, transfer and provenance, of both physical and financial products. Asset management systems require services that can deliver transactional integrity and end-to-end audit support thought-out the lifecycle of the asset. All of this also needs to integrate into legacy systems all of which are exposed to risks associated with asynchronous systems. Privacy issues and the transactional management of access to that data (with immutable proof) is a challenge to the financial services industry, which require:

- Privacy (cryptographically strong data at rest)
- Immutability
- Transactional access authentication
- Privileged data exposure
- Smart contractual exchanges
- Very large IT scalability
- Legacy integration

Title management

Contractual exchange of property and title information currently requires manual processing that requires specialty intermediation including title companies, notaries, and buyer and seller contractual acceptance. A binding immutable ownership ledger provides a universal trail for proof of ownership and exchange history. This capability helps streamline and improved the processes associated with management of any titles such as land, physical property, financial property, and so forth. A title provenance service Distributed Ledger system requires:

- Privacy (cryptographically strong data at rest)
- Immutability
- Transactional access authentication
- Selective data exposure
- Interactive analysis
- Very large IT scalability

Summary

This chapter covered the fundamentals and key concepts of blockchain technology, the factors driving its explosive growth, along with the risks and challenges associated with its adoption. We also provided some detail on specific solutions that could be leveraged to jumpstart your own blockchain journey. Should you have any questions or would like some additional information please contact your HPE representative or select "Contact Us" at the link below.

https://www.hpe.com/us/en/servers/nonstop.html

Index

A

Acquisition models 184, 185
All-flash arrays (AFA) 170, 172, 174
Allocator layer 118
AMD 50–51
Anode 207
Ansible Playbook code 142, 143
Apollo 2200 solution 150–151
Apollo 6500 solution 151
Application-Specific Integrated Circuit (ASIC) 51–52
Application-transparent checkpoint 118
Array Configuration Utility (ACU) 194
Aruba access points (APs) 14–15
Aruba access switches 15
Aruba AirWave 17, 22, 24
Aruba AppRF 17
Aruba Central 22
Aruba ClearPass Policy Manager (CPPM) 17–18
Aruba Instant 21–22
Aruba Instant APs (IAPs) 20–21
Aruba Mobile First campus architecture
   Mobile First Access
      centralized reference design 13, 18–20
      converged wired and wireless endpoints 22–24
      distributed reference design 13, 20–22
      traditional reference design 13–18
   Mobile First Backbone
      collapsed design 25
      deployment 26–27
      hierarchical design 25–26
      VSF 27–29
   sample reference design
      multiple-building 30–31
      single building 29–30
      single VLAN architecture 31–32
Aruba mobility controllers 15–17
Asset management 223
Auto AP detection 23

B

Bill of Materials (BOM)
   DL380 Gen9 management servers 155
   GPU Apollo 2200 154
   HANA TDI workload solution 96
   high-memory (512 GB) Apollo 2200 153
   Management block 133–135
   Mellanox switches 155
   Nimble Storage AF9000 storage array 166–167
   SDX TDI solution 96–97
   Standard Memory (128GB) Apollo 2200 153
   VDI Block 131–133
   XL270d Gen8 8-GPU server 154–155
Bimodal IT 36–37
Bitcoin chart 216, 217
Bitcoin confirmation time 218
Bitcoin signature process 206, 207
BitTorrent technology 207

Index

Blockchain 11
   applications 209
   architectures 219
   asset registration 223
   Bitcoin chart 216, 217
   Bitcoin confirmation time 218
   clearing 212–213
   cost-per-transaction 218
   cryptocurrency 206, 207
   cryptography 205–206
   data integration 222–223
   definition 203
   digital identity 215–216
   distributed ledger 219, 220
   Ethereum 208
   flow 204, 205
   FSI 220
   Hyperledger fabric 207
   MCDLT 221
   participation 209, 210
   payments 213–215
   public, private, and consortium 208–209
   R3 Corda 207, 222
   settlement 212–213, 223
   smart assets 211–212
   smart contract 210–211, 219, 221
   title management 224
Brocade Fabric Vision technology 176
Business 34
Business Intelligence (BI) 158

C

Capex 184, 185
Centralized ledger 212, 213
Centralized reference design 13, 18–20
Clarity dashboard 17
ClearPass Guest 18
ClearPass Onboard 18
ClearPass OnGuard 18
Collapsed Reference Design 13, 25
Commercial NSA (CNSA) suite 46
Composable infrastructure
   architectural design principles 77–78
   IT progression 76–77
   servers 76
Composer 79–80
Consortium model 208–209
Consumption-based IT services
   benefits of 182
   capacity planning 39
   definition 181–182
   elements and attributes 187
   hardware and software components 189–190
   HPE Datacenter Care Service 190
   HPE GreenLake 40
      advantages 186–187
      characteristics 184, 187
      features 183
      flexible capacity 184–185
      flexible technology options 185
      payment 185
      Standard Packages 186
   IT organizations 40
   partial design 188
   phase implementation 190–191
   professional services 40–41
Continuous Integration and Continuous Delivery (CICD) 34–35
Converged Edged Systems
   consolidated architecture 9–11
   IT and OT systems 8–9
Corda SQL 222

Cost per transaction 218
Critical client protection 19–20
Cryptocurrency 206, 207
Cryptography 205–206

D

Datacenter Care Service 190
Data integration 222–223
Data Management Systems (DBMS) 118
Data unification 118
Data Virtualization Platform (DVP) 67–69
DBaaS 38–39
Dennard's Scaling 110
DeployUser 200
DevOps model 34–37
DHCP 24
Digital identity 215–216
Digital signature 206, 207
Disaster recovery (DR) plans 174, 187
Distributed ledger system 207
   vs. centralized ledger 212, 213
   FSI 220
   payment system 214, 215
   public information 219
   title management 224
Distributed Management Task Force (DMTF) 52
Distributed reference design 13, 20–22
Distribution switching nodes 25–26
DL380 Gen9 management servers 155
Docker 144
Docker Swarm Cluster 144
Docker Trusted Registry 144
Dual Inline Memory Modules (DIMMs) 94
Dual Purpose HANA HA configuration 90–91
Dynamic Random Access Memory (DRAM) 47, 66
Dynamic segmentation 19–20

E

Electronic Health Record (EHR) 216
Emulex-Broadcom 174
Encryption 205–206
Enterprise Message Bus (EMS) 120
Enterprise Secure Key Manager (ESKM) 197–198, 200
Enterprise Service Bus (ESB) 122–123
Ethereum 208, 219
Extended Service Pack Overlap Support (ESPOS) 99
EZSwitch 176

F

Fabric-Attached Memory Emulation 119
Factory Express 137
Fault-tolerant programming model 118
Federal Information Processing Standards (FIPS) 140-2 194, 196, 197
Fibre Channel (FC) 169–170
Field Programmable Gate Array (FPGA) processor 66
Financial services 122
Financial Services Industry (FSI) 220
Financial Services Sector 209
Flexible capacity 184–185
Floor space 147
Foresight 119, 120
Frame link modules 83–84

G

10/40 GbE uplinks 22
Gen 6 FC network 171
Gen-Z open systems 114, 115
10 Gigabit Ethernet (GbE) networking 69–70
GitHub 116
Graphical Inference 117
Graphical Processing Units (GPUs) 50–51, 109

Index

Graphical user interface (GUI) 161, 162
Greenfield SAP S4 implementation 103–105
GreenLake 40
    advantages 186–187
    benefits 63
    characteristics 184, 187
    features 63, 183
    flexible capacity 184–185
    flexible technology options 185
    payment 185
    Standard Packages 186

H

HA clustering (PODs) 128–129
HANA Tailored Datacenter Integration (TDI) approach
    business requirements 89
    ECC 93–94
    hardware inventory 96
    memory configurations 94–95
    sample SAP ABAP sizing report 92–93
    SLES for SAP Applications 97–99
    software inventory 97
    Superdome Flex
        features 102–103
        implementation 103–105
        incremental upgrades 103
    Superdome X
        components 91–92
        hardware inventory 96–97
        memory configurations 94–95
        nPartitions 90
        SoH instance 90–91, 99–102
    technical requirements 90
Hard Disk Drives (HDD) 193
High-dimensional search 123
High-network performance 147
High Performance Computing (HPC) 173
    cost 147
    floor space 147
    hardware components 149–150, 153–156
    high-network performance 147
    high-processing performance 147
    project planning implementation 156
    reliable management nodes 147
    server cooling 147–148, 152–153
    server power 147–148, 151–152
    server space 150–151
    solution design 148–149
High-performance memory fabric 114–115
High-performance Storage Area Network (SAN)
    applications 174
    Brocade Fabric Vision 176
    class brocade extension 175
    configuration 171, 172
    deployment services 179
    disaster recovery plans 174
    EZSwitch 176
    Fibre Channel 169–170
    Gen 6 features 171
    HBA 173–174
    HPC 173
    HPE MSA 2052 Storage system 170
    Nimble Storage 170
    NVMe-oFC 172–173
    one-to-one vs. many-to-one deployment 175
    small, medium, and large SAN inventory 177–179
    3PAR StoreServ 170, 176
High-processing performance 147
Hindsight 119, 120
Hierarchical Reference Design 13, 25–26

Host Bus Adapter (HBA) 173–174
Hosted Desktop Infrastructure (HDI) 125
HPE Cloud Technology Partner (CTP) 34
HPE Composable infrastructure
   architectural design principles 77–78
   IT progression 76–77
   servers 76
HPE Composer 79–80
HPE consumption-based IT services
   benefits of 182
   capacity planning 39
   Datacenter Care Service 190
   definition 181–182
   elements and attributes 187
   GreenLake 40
      advantages 186–187
      characteristics 184, 187
      features 183
      flexible capacity 184–185
      flexible technology options 185
      payment 185
      Standard Packages 186
   hardware and software components 189–190
   IT organizations 40
   partial design 188
   phase implementation 190–191
   professional services 40–41
HPE Converged Edged Systems
   consolidated architecture 9–11
   IT and OT systems 8–9
HPE Datacenter Care Service 190
HPE Factory Express 137
HPE GreenLake 40
   advantages 186–187
   benefits 63
   characteristics 184, 187
   features 63, 183
   flexible capacity 184–185
   flexible technology options 185
   payment 185
   Standard Packages 186
HPE GreenLake Flex Capacity Packages 186
HPE GreenLake Standard Consumption Packages 186
HPE HANA Tailored Datacenter Integration (TDI) approach
   business requirements 89
   ECC 93–94
   hardware inventory 96
   memory configurations 94–95
   sample SAP ABAP sizing report 92–93
   SLES for SAP Applications 97–99
   software inventory 97
   Superdome Flex
      features 102–103
      implementation 103–105
      incremental upgrades 103
   Superdome X
      components 91–92
      hardware inventory 96–97
      memory configurations 94–95
      nPartitions 90
      SoH instance 90–91, 99–102
   technical requirements 90
HPE Image Streamer 80–81, 140
HPE InfoSight
   definition 162
   intervolume performance and contention 165, 166
   I/O patterns 163–164
   recurring performance patterns 164, 165
HPE Mission-Critical Distributed Ledger Technology (MCDLT) 221–223

Index

HPE Moonshot 124–125
HPE MSA 2052 Storage system 170
HPE Nimble storage
  AF-9000 solution 159–161
  hardware inventory 166–167
  InfoSight
    definition 162
    intervolume performance and contention 165, 166
    I/O patterns 163–164
    recurring performance patterns 164, 165
  management capabilities 161–162
  OLTP factors
    BI workloads 158
    transactional workloads 157–158
  scaling capabilities 161
  solution requirements 158
HPE NonStop Server 222
HPE Non-Volatile Dual Inline Memory Modules (NVDIMMs) 118
HPE OneSphere
  bimodal IT 36–37
  DBaaS 38–39
  features and benefits 35
  infrastructure options 35
  IT, LOB, and developers collaboration 35–36
  microservices 37–39
  operations, developers, and business executives 39
HPE OneView
  Ansible 141–143
  Docker 144
  hardware and software requirements 141
  Image Streamer 140
  RESTful APIs 139, 140

HPE Persistent Memory NVDIMM 52
  database checkpoint-restore 54–55
  server configuration 53–54
HPE Pointnext 39–40
HPE Pointnext services organization 136–137
HPE Power Advisor 135–136, 152
HPE ProLiant for Microsoft Azure Stack solution
  collective expertise 63
  compliance 61
  connect edge 62
  consumption-based model 63
  data sovereignty 61
  disconnected applications 62
  exceptional speed and performance 63
  hardware inventory 59–61
  maximize performance 61
  modern application development 62
  on-premises data center 58–59
  public and private cloud environment 57–58
  security requirements 61
  unmatched flexibility and configurability 62
HPE ProLiant Gen10
  ASIC enhancements 51–52
  enhanced security
    administrator privileges 45–46
    chassis intrusion detection device 46
    firmware verification 44
    protect, detect, and recover 44–45
    security features and processes 47
    security modes 46
    TPM 47
  GPUs 50–51
  high-speed memory capacity 47–48
  in-server storage density 51

IST
- core boosting 49
- Jitter smoothing 48–49
- workload matching 49–50

OneView 52

performance enhancements 50

Persistent Memory NVDIMM 52
- database checkpoint-restore 54–55
- server configuration 53–54

unified CLI 52

HPE SAN Deployment Service 179

HPE Secure Encryption
- algorithms 196–197
- Broad Encryption Coverage 193
- competitive view 199
- considerations 199
- deployment and management 194
- ESKM 197–198
- flexibility 194
- Gen8 195
- Gen9 195
- Gen10 195
- high availability and scalability 194
- licensing 198–197
- local mode 196
- QuickSpecs 196
- regulatory compliance 194–195
- remote mode 196
- solution overview 195
- system management 200

HPE SimpliVity
- hyperconverged infrastructure
  - appliance 66
  - compute-only scaling 70
  - compute resources 69
  - features and benefits 65
  - federation 66–67
  - hardware inventory 71–73
  - hypervisor clustering 66
  - resiliency 71
  - scale-out workloads 69–70
  - storage efficiency 67–69
- OneSphere 39
- requirements 127–128
- VDI deployment
  - BOM 131–133
  - building blocks 129
  - data center sizing 128–129
  - DL360 Gen10 130
  - DL380 Gen10 130
  - HPE Pointnext Services organization 136–137
  - Management block BOM 133–135
  - rack power requirements 135–136
  - scalability 128

HPE Smart Array (SA) controllers 194, 195

HPE Smart Array SR Secure Encryption 196

HPE Smart SAN 176

HPE Smart Storage Administrator (SSA) 194

HPE Software Defined Infrastructure 139, 144

HPE's Universal IoT platform 9–10

HPE Superdome Flex
- features 102–103
- implementation 103–105
- incremental upgrades 103

HPE Superdome X (SDX)
- components 91–92
- hardware inventory 96–97
- memory configurations 94–95
- nPartitions 90
- SoH instance 90–91, 99–102

Index

HPE Synergy 144
   Composable infrastructure
      architectural design principles 77–78
      IT progression 76–77
      servers 76
   deployment 76
   management ease of use 76
   OneSphere 39
   performance 76
   resiliency 75
   Synergy 12000 frame
      appliance bays 78
      Composer 79–80
      compute modules 81–82
      hardware components 85–87
      Image Streamer 80–81
      networking 83–84
      OneView 78–79
      storage 82–83
HPE System on a Chip (Soc) 124–125
HPE System Update Manager (SUM) 46
HPE 3PAR 82–83
HPE 3PAR StoreServ 170
Hyperledger fabric 207

I

Image Streamer 80–81, 140
Information Technology (IT) 5, 8
InfoSight
   definition 162
   intervolume performance and contention 165, 166
   I/O patterns 163–164
   recurring performance patterns 164, 165
Input/Output (I/O) blender effect 157
Insights 119, 120

Intelligent Edge
   benefits
      bandwidth 5
      business improvements 4
      compliance 6
      cost 5
      duplication and durability 5
      latency 5
      security 5
      supportability 5
   businesses 8
   connect, compute, and control 7
   digital disruption/transformation organizations 8
   elements 3–4
   parameters 8
   real-world information 6–7
   role of cloud 3
Intelligent system tuning (IST)
   core boosting 49
   Jitter smoothing 48–49
   workload matching 49–50
Internet of Things (IoT)
   big data analytics 2
   evolution of computing 2–3
   HPE Converged Edged Systems
      consolidated architecture 9–11
      IT and OT systems 8–9
   Intelligent Edge
      benefits 4–6
      businesses 8
      connect, compute, and control 7
      digital disruption/transformation organizations 8
      elements 3–4
      parameters 8

real-world information 6–7
role of cloud 3
machine learning 2
type of objects 1–2
IT-as-a-Service model 181–182
IT consumption models
application architectures 109
data revolution 111–112
Dennard's Scaling 110
GPUs 109
properties governing scaling 109, 110
System of Action 108
System of Engagement 108
System of Record 108

K

Key Management interoperability Protocol (KMIP) 196
Key Performance Indicators (KPIs) 101
Kubernetes 37–38

L

Licensing 198–197
Lines of Business (LOBs) 33, 35–36
Linux 116, 118
Local mode 196
Logistic systems 120–122

M

Managed Data Structures (MDS) 117
MDC developer toolkit
data management 117–118
emulation and simulation tools 119
GitHub 116
Graphical Inference 117
image searching 117
Linux 118
persistent memory toolkit 118
programming frameworks 117–118
MDC software stack 116
Mellanox switches 155
Memory-Driven Computing (MDC)
advantage 112
attributes 113
definition 107
developer toolkit
  data management 117–118
  emulation and simulation tools 119
  GitHub 116
  Graphical Inference 117
  image searching 117
  Linux 118
  persistent memory toolkit 118
  programming frameworks 117–118
ESB 122–123
financial services 122
high-performance memory fabric 114–115
IT consumption models
  application architectures 109
  data revolution 111–112
  Dennard's Scaling 110
  GPUs 109
  properties governing scaling 109, 110
  System of Action 108
  System of Engagement 108
  System of Record 108
Moonshot 124–125
vs. processor-centric computing 112
shared persistent scale-out memory 114
similarity search 123–124
SoC 124–125
software-defined composability 115
software stack 116

task-specific processing 115

time machine 119–120

transportation and logistics 120–122

Mission-Critical Distributed Ledger Technology (MCDLT) 221–223

Mobile First Access

centralized reference design 13, 18–20

converged wired and wireless endpoints

  AirWave integration 24

  AP integration 22–23

  ClearPass integration 23–24

  features 22

  ZTP 24

distributed reference design 13, 20–22

traditional reference design 13

  Aruba access switches 15

  Aruba AirWave 17

  Aruba APs 14–15

  Aruba CPPM 17–18

  Aruba mobility controllers 15–17

Mobile First Backbone

collapsed design 25

deployment 26–27

hierarchical design 25–26

VSF 27–29

Moonshot 124–125

MSA 2052 Storage system 170

MSRUser 200

Multi-Process Garbage Collector (MPGC) 117

N

National Institute of Standards and Technology (NIST) 194

NetWeaver Virus Scan Interface (NW-VSI) 99

Network File System (NFS) 70

Nimble Connection Manager (NCM) 161

Nimble storage

AF-9000 solution 159–161

hardware inventory 166–167

InfoSight

  definition 162

  intervolume performance and contention 165, 166

  I/O patterns 163–164

  recurring performance patterns 164, 165

management capabilities 161–162

OLTP factors

  BI workloads 158

  transactional workloads 157–158

scaling capabilities 161

solution requirements 158

Noisy neighbor workloads 165, 166

NonStop Server 222

Non-Volatile Memory (NVM) 118

Non-volatile memory DIMM (NV-DIMM) 47

Non-Volatile Memory Manager (NVMM) 118

nPartitions 90

NVMe over Fiber Channel (NVMe-oFC) 172–173

NVM latency and bandwidth 119

O

OneSphere

bimodal IT 36–37

DBaaS 38–39

features and benefits 35

infrastructure options 35

IT, LOB, and developers collaboration 35–36

microservices 37–39

operations, developers, and business executives 39

OneView 52

Ansible 141–143

Docker 144

hardware and software requirements 141
Image Streamer 140
RESTful APIs 139, 140
Synergy 12000 frame 78–79
On-Line Analytical Processing (OLAP) 157–158
On-Line Transaction Processing (OLTP) 157–158
Operational Technology (OT) 5, 9
Opex 184, 185
Oracle operation size heatmap 163
Oracle read-write size heatmap 164
Oracle recurring performance patterns 164, 165
Order Management (OM) 103–105

P

Path Selection Policy (PSP) 161
Payments 213–215
Performance emulation 119
Per-Port Tunneled Node (PPTN) basis 19
Persistent Memory NVDIMM 52
database checkpoint-restore 54–55
server configuration 53–54
Persistent memory toolkit 118
Per-User Tunneled Node (PUTN) basis 19
PoE devices 23
Pointnext 39–40
Pointnext services organization 136–137
Power Advisor 135–136, 152
Power on Self-Test (POST) 52
Private model 208–209
Processor-centric computing 112
Production Mode 46
Projects 35–36
ProLiant for Microsoft Azure Stack solution
collective expertise 63
compliance 61
connect edge 62
consumption-based model 63
data sovereignty 61
disconnected applications 62
exceptional speed and performance 63
hardware inventory 59–61
maximize performance 61
modern application development 62
on-premises data center 58–59
public and private cloud environment 57–58
security requirements 61
unmatched flexibility and configurability 62
ProLiant Gen10 149
ASIC enhancements 51–52
enhanced security
administrator privileges 45–46
chassis intrusion detection device 46
firmware verification 44
protect, detect, and recover 44–45
security features and processes 47
security modes 46
TPM 47
GPUs 50–51
high-speed memory capacity 47–48
in-server storage density 51
IST
core boosting 49
Jitter smoothing 48–49
workload matching 49–50
OneView 52
performance enhancements 50
Persistent Memory NVDIMM 52
database checkpoint-restore 54–55
server configuration 53–54
unified CLI 52
Public-key cryptography 206
Public model 208–209

Q

Quality Assurance (QA) SoH instance 90–91

R

Radix Tree 118
Recurring performance patterns 164, 165
Redundant Array of Inexpensive Disk (RAID) 71, 129, 194
Redundant Array of Inexpensive Node (RAIN) 71
Reliable management nodes 147
Remote mode 196, 200
Representational State Transfer (REST) Application Programming Interface (API) 139, 140
Rogue AP isolation 23
R3 Corda 207, 221, 222

S

SAN Design Reference Guide 171
SAP Enterprise Central Component (ECC) 89, 93–94
SAP's Suite on HANA (SoH)
   deployment
      implementation plan 99–100
      overview 101–102
      testing 101
      ECC production environment 93–94
      Quality Assurance (QA) 90–91
Scalable Persistent Memory (PMEM) 48
Scale-out group 161
Scale-out memory, MDC 122
Scale-to-fit design 161
Secure Encryption
   algorithms 196–197
   Broad Encryption Coverage 193
   competitive view 199
   considerations 199
   deployment and management 194
   ESKM 197–198
   flexibility 194
   Gen8 195
   Gen9 195
   Gen10 195
   high availability and scalability 194
   licensing 198–197
   local mode 196
   QuickSpecs 196
   regulatory compliance 194–195
   remote mode 196
   solution overview 195
   system management 200
Self-Encrypting Drives (SEDs) 193, 199
Server cooling 152–153
Server power 151–152
Server Profile Template 142
Server space 150–151
Service Level Agreement (SLA) 37–38
Service Oriented Architecture (SOA) 34–35
Shared persistent scale-out memory 114
Shoveller 118
Similarity search 123–124
SimpliVity
   hyperconverged infrastructure
      appliance 66
      compute-only scaling 70
      compute resources 69
      features and benefits 65
      federation 66–67
      hardware inventory 71–73
      hypervisor clustering 66
      resiliency 71
      scale-out workloads 69–70
      storage efficiency 67–69

OneSphere 39
requirements 127–128
VDI deployment
  BOM 131–133
  building blocks 129
  data center sizing 128–129
  DL360 Gen10 130
  DL380 Gen10 130
  HPE Pointnext Services organization 136–137
  Management block BOM 133–135
  rack power requirements 135–136
  scalability 128

Smart Array (SA) controllers 194, 195
Smart assets 211–212
Smart contract 210–211, 219, 221
Smart Rate ports 22
Smart Storage Administrator (SSA) 194, 200
Software-defined composability 115
Software-Defined Data Center (SDDC)
  API-driven technology 34–35
  DevOps model 34–35
  digital transformation 33–34
  HPE consumption-based services
    capacity planning 39
    HPE GreenLake 40
    IT organizations 40
    professional services 40–41
  HPE OneSphere
    bimodal IT 36–37
    DBaaS 38–39
    features and benefits 35
    infrastructure options 35
    IT, LOB, and developers collaboration 35–36
    microservices 37–39
    operations, developers, and business executives 39

Software Defined Infrastructure 39, 139, 144
Solid State Drives (SDD) 193
Spark 117
S4 Supply Chain (SC) 103–105
Statement of Work (SOW) 190
Storage Area Network (SAN)
  applications 174
  Brocade Fabric Vision 176
  class brocade extension 175
  configuration 171, 172
  deployment services 179
  disaster recovery plans 174
  EZSwitch 176
  Fibre Channel 169–170
  Gen 6 features 171
  HBA 173–174
  HPC 173
  HPE MSA 2052 Storage system 170
  Nimble Storage 170
  NVMe-oFC 172–173
  one-to-one vs. many-to-one deployment 175
  small, medium, and large SAN inventory 177–179
  3PAR StoreServ 170, 176

SuiteB 46
Superdome Flex
  features 102–103
  implementation 103–105
  incremental upgrades 103

Superdome X (SDX) TDI solution
  components 91–92
  hardware inventory 96–97
  memory configurations 94–95
  nPartitions 90
  SoH instance 90–91, 99–102

Index

SUSE Linux Enterprise Server (SLES) 97–99
Symmetric-key cryptography 205
Synergy
    deployment 76
    HPE Composable infrastructure
        architectural design principles 77–78
        IT progression 76–77
        servers 76
    management ease of use 76
    OneSphere 39
    performance 76
    resiliency 75
    Synergy 12000 frame
        appliance bays 78
        Composer 79–80
        compute modules 81–82
        hardware components 85–87
        Image Streamer 80–81
        networking 83–84
        OneView 78–79
        storage 82–83
Synergy 12000 frame
    appliance bays 78
    Composer 79–80
    compute modules 81–82
    hardware components 85–87
    Image Streamer 80–81
    networking 83–84
    OneView 78–79
    storage 82–83
System of Action model 108
System of Engagement model 108
System of Record model 108
System on a Chip (Soc) 124–125
System Update Manager (SUM) 46

T

Task-specific processing 115
Technology 34
Thermal Design Point (TDP) 49
Third-Generation Partnership Project (3GPP) 5
3PAR StoreServ 176
Time machine 119–120
Title management 224
Top-of-Rack (ToR) switches 136, 177, 178
Trade Agreement Act (TAA) 47
Traditional reference design 13
    Aruba access switches 15
    Aruba AirWave 17
    Aruba APs 14–15
    Aruba CPPM 17–18
    Aruba mobility controllers 15–17
Transactional workloads 157–158
Transportation 120–122
Trusted Platform Module (TPM) 47

U

Unified CLI 52
Unified policy enforcement 20
Universal Control Plane 144

V

Validated Reference Design (VRD) 32
Very Small Aperture Terminal (VSAT) 5
Virtual controller (VC) 21–22
Virtual Desktop Infrastructure (VDI) deployment
    Block BOM 131–133
    building blocks 129
    data center sizing 128–129
    DL360 Gen10 130

DL380 Gen10 130
HPE Pointnext Services organization 136–137
Management block BOM 133–135
rack power requirements 135–136
scalability 128
Virtual Local Area Network (VLAN) 31–32, 141
Virtual Machines (VMs) 50
Virtual Switching Framework (VSF) 27–29
VisualRF 17
vSphere Web Client 66–67

W

Wide Area Network (WAN) 5
WLAN 31–32
Write-Ahead-Logging Library (WAL) 118

X

XL270d Gen8 8-GPU server 154–155

Z

Zero-Touch Provisioning (ZTP) 24